THE
P'TAAH TAPES

Transmissions From The Pleiades

AN ACT
OF FAITH

Channelled by Jani King

TRIAD
PUBLISHERS PTY LTD

For information address:

TRIAD Publishers Pty. Ltd.
P.O. Box 731, Cairns
Qld 4870
Australia

Book Title

AN ACT OF FAITH
Transmissions From The Pleiades

Author: Channelled by Jani King
Edited and narrated by Peter Erbe
Book and cover design: Peter Erbe
Typeset by Thomas Williams

National Library Of Australia: ISBN: 0 646 07469 5
Printed by Kaleidoscope, Perth, W.A., Australia

Triad Publications aim at aiding and inspiring
a spiritually unfolding humanity.

Who you are is wondrous.
Who you are is beauty beyond
measure.
Who you are is the Source expressing
in this density of reality.
Soon you will know that you are the
GOD I AM.
It is called coming home.

– P'TAAH –

You have created yourselves into this incarnation, that you may bear witness and that you may contribute to the changes which are coming, the changes to occur on your planet. You do not have so much time to wait – it is imminent.

For each one of you there is a yearning within your breast to experience these changes, to experience the change in the consciousness of humanity; the change within your Earth itself. And she indeed is quickened.

The words you have will not describe the ecstatic explosion, nor can you imagine how you will all be in that time, when every atom and molecule upon this planet, and the whole planet herself, will radiate with divine light. Such exquisite beauty is beyond imaginings. And when the transition has taken place there will be beings flocking to your planet to sing the wildest hosannas, in joy, in jubilation, in thanks, in blessings. And those beings will be the unseen ones, they will be the beings from other worlds who do not appear as you do, and yet you will be able to perceive the divinity in all things. It will not matter the shape and the size, you will perceive the Godhood in all. It will be most magical.

– P'TAAH –

ACKNOWLEDGMENTS

My heartfelt thanks to Peter and Carol Erbe for
your loving support, for allowing us to gather in
your beautiful home for all those months, and for
the painstaking and patient work in transcribing
and editing P'taah's material.

To all the lovely and loving people who have
supported me during the making of this book,
thank you.

To David John Ward: None of this would have
been possible without you, without you bringing
Germain to Australia and thus changing the lives
of all of us. My grateful thanks.

Jani King
Kuranda
December, 1991

CONTENTS

FOREWORD

By Peter O. Erbe

Something highly unusual, and very special, took place in the latter half of the year 1991. Amidst the tropical setting of the coastal hills of North Queensland, Australia, a group of people gathered regularly for sixteen weeks to listen to the teachings of one of the star-people: P'taah from the constellation of the Pleiades.

Since time immemorial, Man has been driven in the search for knowledge, which in truth is but the drive for Self-knowledge, the desire for spiritual identity. This desire, as history well documents, exceeds even the need for self-preservation. Unfortunately, whenever a member of the human family found the priceless pearl and felt compelled to impart this knowledge, we worshipped the messenger, elevated him to Godhood and not only institutionalized the message, but veiled it with much hocus-pocus as well. This placed both, message and messenger, at a safe distance from our daily lives and ensured - of necessity - that we could be thoroughly controlled. For in this manner we freed ourselves from the yoke of having to take responsibility; we gave it to the intermediaries, the keepers and caretakers of religions and dogmas - the priests, the gurus, the popes. What an ingeniously simple method to give one's power away. If there be any among us who believe that things have somewhat improved, then let them be reminded that what appears as improvement is but a shift from surrendering power to religious leaders and organizations to - for one - handing it over to governments or political systems. The spectrum of those who wield power has become broader, that is all. The inclination of the individual to unload his responsibility remains the same, but so does the cry of the human heart for true liberation; and it has been heard.

It is credit to the star people and our brothers in the unseen for having devised a procedure for imparting knowledge without leaving any feet to be kissed. Thus we have entered the era of channelling, a most intriguing concept for those not familiar with it, for there are no pictures and symbols to be hung on the wall, no sacred halls wherein idols can be worshipped. On the contrary, for here the only temple to be entered is the sanctum within and there we inevitably discover our Being, the God I Am, as the Source of all power and thus reclaim our heritage. In other words, to benefit from this knowledge we have to actually face ourselves - we have to take responsibility.

For those who find the courage to do just that, the difficulty in profiting from channelled teachings lies not so much with the concept of channelling itself, but with the genuine consideration of the thought conveyed by the particular entity being channelled. For here is, where a real clash of 'realities' can occur. Whenever we are confronted with thought which does not fit the matrix of our individual or group 'reality', it is most likely regarded as 'unrealistic'; much in the way a South African Bushman would exercise great suspicion, if being acquainted by a Harvard professor with the theory of quantum physics. All the more should we honour the patience and love displayed by certain channelled beings.

At a time of a humanity stressed to the hilt and suffering as never before from the harsh effects of an unremitting, fear-based consciousness, the denial of the God within, P'taah's communication could not have been timed more appropriately.

P'taah prepares humanity for the forthcoming transition from separation to Oneness. If there ever was a message of limitless love, of joy and upliftment, of concrete thought applicable to practical, every-day living, then it is this material; though much of it shakes the bed-rock of belief structures which constitute human reality.

P'taah tells of the grand changes to come for humanity and the planet Earth, he opens our vista to a universe teeming with life, he speaks of the inner-Earth people and the star-people, and in doing so, assures us that we are not alone. What is more, he presents us with the panacea to transmute fear into love, to discover who we really are. Gently he dissolves the imprisoning shackles of dogma and concept, which lock Man into a consciousness of survival thinking, and reveals, contrary to all appearances, an irresistible, breathtakingly beautiful destiny for Mankind.

The love of the star-people for humanity could not be expressed any more than in P'taah's own words:

"WE WILL DO ANYTHING TO BRING YOU HOME!"

Peter O. Erbe
December 1991

AN INTRODUCTION TO JANI KING

It is conceivable that many a reader might be desirous of acquainting him or herself to a certain extent with Jani King, the channel, although usually, with the process of channelling, the entire focus is on the entity being channelled and one can easily become forgetful of the fact that here is a living human being, dedicating his or her time and energy to making it all possible.

In Jani's case we are facing a warm-hearted, down-to-earth woman who loves nature, who loves to laugh and who, we might add, has her own brand of humour - and lots of it, too.

So, to acquaint ourselves with Jani, a question and answering session with Jani and the group was arranged, during which Jani could answer the most pressing questions herself.

Q: Jani, I would like to know how you and P'taah met.
Jani: Okay. I have to start off laterally rather than formally for this. In 1947 I lived in New Zealand - I was born there. I lived in a very remote area, in the middle of 350 square miles of pine forest. Bearing in mind that it was just after the war, there was no transport; so it was really very, very isolated. So, at the time I was a very little girl and I went wandering from our house into the forest and I remember very clearly going into the forest and then something occurred. I had no idea what it was and then sometime later, when I came back to the house, it seemed much later in the day. Extraordinarily enough nobody had realized that I had gone. For those of you who are mothers of little girls, if they are gone for just an hour, you start to become very worried. Yet nobody seemed to have noticed that I had gone, in spite of the fact that it had been for many hours.

1

Sometime during 1988 a channelled energy named St. Germain had come to Australia. During this time I had picked up a book by Whitley Streiber called 'Communion'. When I saw the book on a friend's coffee table I experienced the most extraordinary feeling of being drawn to it. I asked my friend what the book was about and he answered: 'Some stuff about a guy being abducted by aliens'. The cover of the book showed this triangular face with huge eyes and for some reason I reacted with goose bumps and I felt very emotional without having any idea what caused it. I held the book in my hands and had to force myself to put that book down. The next day I locked up the house, disconnected the phone, sat down and read the book from cover to cover. It seemed so familiar to me, except for the fact that Whitley Streiber was terrified when he came to understand what his ordeal was. I did not have that feeling at all. 'Coincidentally', a couple of days after that, I spoke to St. Germain and I do not know why I said it, but I asked: Could you tell me about the experience I had in 1947 (as a child), that I do not remember. St. Germain said: 'You were taken aboard a craft. There was a medical procedure performed to put in place certain knowledge and memory that will come to the surface in the next now moment of your time'. He went on to describe, what this abduction was about. I sat there, going 'oh, yeah, yeah, yeah' with my eyes rolling in my head. And yet it seemed that when he finished and I had assimilated what he had said, it was not a surprise to me.

I also asked him about a visit I had from another being, and I am not talking about an apparition but about physical reality. It was in 1961 and it was P'taah who came in a physical body (I do not know how many physical bodies he has), and we had a communication, which was quite, quite extraordinary. That is the only time in my conscious, waking knowledge, that I have met P'taah. I had hypno-regression about the early experience in 1947 and now I do have conscious memory of what occurred then. The experience in 1961 in a way was so extraordinary, it has left me still with a residue of grief. *(Recalling this experience, Jani is deeply moved and pauses for a moment to regain her composure.)* And this is because - really - I am still waiting.

I have to say to you that my experience with P'taah from day to day is very different to yours. For you it is almost....I cannot explain it - you have a different experience than I have. I can talk to P'taah in my head. He talks to me, but it is much more ephemeral than your one to one experience with him.

Q: Have you asked P'taah about this feeling of grief?

Jani: Yes, I have.

Q: What did he say, if I may ask?

Jani: Well, I guess he says the same thing to me as he says to you about grief; that it is all transmutable. In a way I have dealt with it, because for a long time I was really waiting - I mean waiting day by day. Now I am not and it is okay if I never come face to face with him again, but there is still this thing there in my heart, you know.

Q: When P'taah comes through, there seems to be an emotion and your eyes become slightly moist. To what extent is he affected by emotions? I would presume it is different with him than with us. Have you any knowledge about that?

Jani: I think you should ask *him* about that. All I can say is: not only do my eyes leak, so does my nose. We have had quite a chat about this, the 'boss' and I, and I said that really I have no control over what he does with my body, but I did ask him not to make it too 'disgusting'; like walking around with a runny nose. He is pretty good but very often when I come back my nose is almost, but not quite, running. The other thing is often my eyes tear, although there is no emotion involved for me in the sessions.

Q: Jani, where do you go, when P'taah comes in?

Jani: I have no idea. Sometimes it is sort of as if I am about here - not very far away. *(Jani points to the right and back of her head.)* Sometimes I have a sense of what has been going on, sometimes I have a recall of some things that he has said, yet I do not have any recall of what anybody else has said. Sometimes I get very smart. The other day Peter did a session with P'taah and when I came back I said to Peter "Oh yeah, I got that. I know what that was all about'. However, when Peter told me about the session, I really only knew a very small part of what had been said, so that just shows you how smart I am. The answer really is that I have no idea. Sometimes, and I don't know whether its got anything to do with what he (P'taah) says, or what he is doing in the sessions, I really don't want to come back. For those who were here last week, it was a very emotional thing, and I don't know why that was, but I didn't want to come back. When I was typing up the transcript I didn't find anything in there that I would really get that emotional about. All I knew was that I really would have preferred not to be here.

Q: Jani, it's about leaving your body. The first time, how did you let yourself go?

Jani: I didn't. It was involuntary.

Q: How do you mean? Do you mean he just came and pushed you out?

Jani: No, no. I mean I didn't really understand what was happening. The first time it happened it was not as if I had gone anywhere, it was that I heard this voice that sounded as if it was coming out of a cave, coming out of my mouth. I said 'Oh, shit. What is happening?' But I understood very quickly what was going on. The other thing is that P'taah does 'pop in'. I know that he is there, and it is never ever without my permission, so to speak, that he comes in. When it is a session like this, I actually go through a little exercise, which is what I do when I sit here with my eyes closed, going mumble-mumble-mumble, then I leave and he comes in. But sometimes it can be as quick as the blink of an eye, which can be hysterical, and can cause quite a furore, particularly when people are in the group and who have no idea what is actually happening. One minute I am there and the next minute there is someone else talking to them. A bit disconcerting. It happens mostly, that involuntary stuff - I say involuntary, because I do not go through all the mumbo jumbo to get out - if I'm very relaxed, if I am having a good time, especially if I have had a couple of glasses of wine and feel really laid back. The next minute 'whacko' here he comes. Usually because he wants a glass of wine, too. Let there be no mistake about that. *(Gales of laughter.)*

4

Q: Jani, when you said P'taah appeared in a physical form, do you mean as a man or woman, or something different? Something we wouldn't be familiar with?

Jani: Something different. It was a light body, but a physical body. Not as dense as ours. A humanoid form. He had very beautiful, very large eyes. That was the most incredible thing. I can't really tell you too much about it, because what was happening was like being programmed. Like a computer receiving all sorts of data. But the other thing concerned my heart. I was being inundated with this most incredible love. Unlike anything else I had ever experienced. Since then I have experienced the same feeling with Saint Germain - channelled through Azena Ramanda - and also while swimming with whales in the wild.

Q: Does P'taah communicate with you about what you, as an individual, should be doing for your own expansion?

Jani: He puts little gem-like jewels in my ear, and a great kick up the bum, when I am not doing as it could be done for my greatest (spiritual) benefit.

Q: Does it tax you physically?

Jani: It depends. if I am feeling really on top and full of energy, after a session I feel very energized and that energy lasts for hours. If I am tired and not feeling very bright before a session, I feel energised when I come back, but when that 'wears off' I am exhausted.

Q: When people have a personal session with P'taah, does he know all about that person? The beliefs and the past and all of that stuff?

Jani: I don't know. Who's had a personal here?

Peter Erbe: I have. Can I answer?

Jani: Sure.

Peter: He knows what colour undies you are wearing. (Gales of laughter.) Very personal.

Q: Every so often in the sessions we hear about scales. What is that about?

5

Jani: It's a joke. After a very heavy, emotional three-day workshop in Melbourne last year, P'taah was talking to the participants. The feeling among the people was very highly charged. A real high. P'taah told them that if he was in his own body he would give them a light-show to reflect to everybody there how exquisitely beautiful they were, and what beautiful lights they were radiating. When he said this, the electric lights in the room flickered on and off a few times. Someone asked him 'What do you look like?' P'taah answered 'Oh, I'm very beautiful'. The facilitator of the workshop muttered out of the corner of his mouth to P'taah 'Yeah, if you don't mind green scales'. Everyone shrieked with laughter, and P'taah turned to this friend and answered 'Ah, but beloved, you cannot see the scales for the light'. So that has sort of become a standing joke with P'taah.

Q: Have you asked to go there, to the Pleiades?

Jani: Apparently, I have been. He (P'taah) says there have been many times over the last thirty years when we have been together - when I have been off-world, been to the Pleiades. I have to tell you, I have no conscious memory of it, although I have memory of being on a craft, apart from when I was a child. I cannot tell you much about it, though. It doesn't really make sense within the logical way we perceive things. I keep saying that I really would like to remember these other experiences I'm supposed to have had, but he keeps saying that when the time is right I shall have full memory of everything. He says the time is coming soon, although what 'soon' means as far as they are concerned, God knows. I try not to hang out for it, you know. It is a bit difficult sometimes. Sometimes something sparks me off and I get this intense yearning, but most of the time I am pretty au fait with it. I cruise along with it and think when the time is right, it's right. I trust P'taah and my own soul integrity in the knowing that whatever and however it all happens, it will be for my own highest good. I have the doubts and the low times and wonder what the hell I am here for, the same as everybody else, but you know, when you come right down to it, I love him, and there's nothing else I'd rather be doing than living right here and now. I guess I'd have to say I am one helluva happy person, most of the time.

6

Chapter 1

FIRST TRANSMISSION.
Date: 29th of August, 1991.

P'taah: Greetings, dear ones. That which is humanity is indeed star-seeded. Eons before your time, before the time of your history, the star people have seeded your planet. And indeed the star people have never been far from humanity - always in these thousands upon thousands of years there have been many comings and goings. In your ancient books there is much writing which has become what is called mythology - about these grand meetings with that which was human of that time and the star people, who were elevated to Godhood. Indeed, the star people are not God-hood any more than you are. The only difference really at this time is that the star people, indeed, know that they are expressions of divinity, and humanity has forgotten. The humanity of this planet lives in what is called separation of Self from SELF. You have forgotten that every facet of who you are is indeed an expression of divinity.

Now, many of humanity in this time believe that this physical life is indeed your, what is called, 'one shot' at it. It is not so. Reincarnation is indeed a fact, whether or not you believe it. Now, some of the other religions indeed do talk about reincarnation, but there is much dogma attached to this belief that is not in line with what is called universal truth. Now, we will touch on that briefly, because it is important that you understand. In many of what you understand to be eastern religions, which talk of reincarnation, there is a thought, that what you do today you get punished for tomorrow and we are speaking in lifetimes, not in actual days. It is not so. When you think of punishment you are thinking of judgment and indeed there is no judgment of who and what you are outside of your own mind.

There is no such thing as a universal judgment. It is important that you understand this, because as we speak further unto you, you will come into an understanding that one of the things which cause the separation of Self from SELF is judgment, indeed the very harsh judgment you have of Self.

Now, reincarnation - the wheel of life some of your ancient religions name it - it is indeed rather like a wheel, however, we may say it like this: There is that which is called soul energy and the soul energy is indeed part of the knowing of its own divinity and with each incarnation, with each lifetime aspects of the persona - if you like - goes back to the Source and the thread continues on to the next lifetime. We have noticed a quite amusing characteristic, when people are becoming involved in what is called past lives, hm? Past lives - everybody is very interested in who and what they were. Now, the truth is, you have been everything. Every facet of every human you are, whether or not you understand this. Of course, humans are very happy when somebody would tell them that they have been Cleopatra or Napoleon, hm? And then somebody will say: 'But it is impossible because indeed I was Cleopatra'. Now, the truth is that there are probably thousands, who are part of the soul energy of the entity known as Cleopatra. Do you understand? The aspects return to the Source and as they come forth again, so it is dispersed and as the wheel continues, so there are more and more and more. And ultimately, dear ones, all are part of One. *There is no separation.* You are everything that has ever existed, that is what you are.

Now, when humans are birthed, and let us speak of this moment in time, as they are birthed and come into the persona that they have chosen and the family situation which they have chosen, so they are already integrating within them those aspects of all, all of humanity. What you are, indeed, is looking outside of yourself to reflections of who you are. When you are still a babe you tap into what is called the collective consciousness of humanity and so you are coming into a belief structure, which is common to your culture, that which is your country, that which is your race. And as you know full well, it varies from country to country and indeed from parts of a country to parts of a country. What is common to all is judgment, the invalidation of you as a child and it is from this judgment and invalidation of who you are, that you learn judgment, that you learn to invalidate who you are. And so, by the time you come to your adult years it is already set in concrete; you have started to invalidate that which you be, that is why it is so painful for you to look at who you are and so you project the judgment outside of yourself. It is for you to know truly, that everything that exists outside of yourself is merely a reflection of that who you be. *There is no facet of human behaviour that you have not been.* You have been what is termed victim and you have been what is termed perpetrator. It is important for you to understand that *you create your own reality absolutely.* That who you are is grand, powerful energy, but you do not know your own power, you do have no idea what wondrous, creative, multidimensional entities you truly are. But dear ones, now is the time for you to come into this understanding. This time is a time of great change. There is the cyclic change of your Earth, there is also the change of consciousness of humanity, a time which will be fear-filled for many, but it is for you to know that, indeed, you are creating everything and you may change everything. You may create the changes with fear or with love. If you create the changes with fear you will come to know the desolation and devastation of your planet. If you truly are in understanding that it is you who creates reality, then it is you, indeed, who may create a wondrous paradise, that if you go forth with courage, with love in your heart, then, indeed, there is no need for the devastation and you will certainly - as time goes by - understand that no thing is set in concrete.

When we speak of multidimensional we are not merely speaking of those realms that you cannot consciously come to, that you would think of as other time zones, etcetera; we are speaking of the dimensions of this your Earth. You see, there is not merely one Earth. There is not merely one life. There are not merely unchangeable situations. There are virtually as many dimensions as there are people, and each one of you is the central sun of your universe. Each one of you creates your own universe, indeed co-create with those you draw into your life and you may have everything; there is abundance for all upon your planet, there is everything you could possibly desire there for you.

The art of manifesting, dear ones, is one that you truly know very well, it is just that you do not understand how it operates, but you do it every day. Manifesting, hm? You can create abundance, you can create joy, wondrous laughter. You can create a world of love - hm? *You may have it all.* It is also for you to understand, beloved ones, that there are only two expressions truly upon your plane. One is love and all that wondrousness that goes with it and *everything which is not an expression of love is indeed an expression of fear;* and always, always you may choose. And so, when your opportunities arise you must understand that you may choose love or you may choose fear. Now, we would at this time wish that you may start to ask questions.

Q: (F) We are all one. That means that you and everything that we can even think of which exists, is all part of the same Oneness. Why are we limited and you not?

P'taah: Now, it is so that we are all One, because we are all created from the Source, indeed? The only difference is that we understand how the universe works and you do not consciously. But, you see, in truth all knowledge, all knowing is within each and every one of you. Each and every one of you is busy hurrying, scurrying around searching, searching for enlightenment. Dear ones, every word, every piece of knowledge that has been given forth is more than enough, because all of your writings from all of your teachers during eons of time tell you the same thing. Everything is within. Now, as you become more and more expanded in consciousness it is as if you were having a radio antennae within you and as you are coming more and more into expanded awareness, so it is as if the antennae becomes broader and broader to receive more and more. And as this occurs, the computer within your head becomes more and more facilitated. Do you understand? So it is not that you should go out searching - it is not necessary. The best of teachers you may have is within your nature. And indeed, as you become more and more engrossed in that which is natural heritage upon your planet, that which is your garden, that which are your creatures, that which is your sea, your sky, your seasons, so you come more and more into the rhythm of this your world - that which is your *real* world.

It is also that as you become more and more in tune with your body, there is less separation between the body and the mind - you will understand that your physical body, the structure of your body, is indeed a microcosm of your universe. Your body, indeed, speaks to you. When there is a disease within the body, it is only a reflection of the dis-ease emotionally, *because all in physical reality is reflection - all*. And as you become more and more in tune, so you will understand your body more and more. Now, as you begin to tap into your own knowing, so you become less and less separated of Self from SELF, you understand?; more and more integrated with the total SELF, indeed. Is this answering your question, dear one?

Q: It gives me plenty to think about. Thank you.

P'taah: Question?

Q: (M) What can we do to become more attuned to this knowing, what can we do to hasten up? I am impatient.

P'taah: Indeed, then it is to go slowly.

Q: Is there not any way to be more active concerning the expansion of consciousness?

P'taah: Indeed, now, we would say, the more you are searching, the more you are hurry-scurrying, the slower is the progress. You see, dear one, it is a grand dichotomy, like everything on your plane. It is going with the flow of that which is life. It is to be. The greatest aligner of all is laughter. Play more, laugh more. It is true, you know. It is not do's and don'ts. You know, there is no such thing as right and wrong, there is no such thing as good and bad, really. And in truth there is no such thing as a wrong decision. Everything is a learning process, and that is why you are here, dear one. It isn't enlightenment. The more you search, the less you will find. Our woman[1] has a very delicious saying - we are here for a good time, not for a long time. And it may be a good time. Much love and laughter, much dancing naked under the Moon, that is what we prefer. To be delicious amongst your nature, to be in hearty fellowship - not to worry about

---

[1] P'taah refers to Jani King as 'our woman'.

11

enlightenment. As you worry about it, beloved, it will become further and further away. It is to know, truly, that you are divine expression, no matter what. And in truth you are already enlightened beings - you just have forgotten. No good or bad - no right or wrong, no judgment. No judgment, dear ones. *There is no God up there keeping a note about what you are doing.* We say very often, that humanity grew with the conception of a God who is an old man, who sits in the clouds. And even when you are grown and intellectually you have dismissed this mythology - you still, in your hearts - all of you - think that there is God up there in a cloud keeping tabs on you. Well dear ones, I have hurry-scurried all over the sky and I have looked and looked and I have not found him. *That which is God is who you are. You are grand God/Goddess.* That is who you are in truth. That is who you are.

> **Q: (M) But we suffer amnesia!**

P'taah: Indeed. But it is the merest veil. Now, you see, all that you need to do is to be as the beloved creature there, who is not worried about God at all, but is having a perfectly wonderful evening. *(P'taah points at the house-cat curled up blissfully in the lap of the hostess).* Questions?

> **Q: (M) When you say there is no good or bad, how do you plot your course for the future?**

P'taah: Without judgment, dear one.

> **Q: So, if there be two roads to follow, judge neither, but follow which one?**

P'taah: Do what makes your heart sing. You see, where you are coming from - a place of love - how can there be a wrong decision - hm? It does not matter. Do what makes the heart sing for the moment. You may decide, a little further along, that it is your heart's desire to do something else, then you may. It is not that you must make a decision forever. Forever is a long time, beloved. And always you may change your mind. Now, we are in understanding - certainly - that in your culture you are expected to make certain decisions in your life and to have to live by them forever. Well, it does seem - hm - a trifle silly, does it not? It is not necessary. You may change whenever you wish, and that is alright. Now, we are not speaking in any way of social consciousness. And for many of you, to do that which makes your heart sing, may fly in the face of that which is social consciousness and then you must decide which is more important, your own joy and happiness or that which is for society. There is no right or wrong. Only in the mind of humans. Now, let us not dismiss discernment. When we speak of non-judgment we are not speaking of what is called lack of common sense. You may be discerning and you may say: 'I would prefer this or that.' Judgment is indeed, when you make something 'right' or 'wrong', 'good' or 'bad'. Do you understand what I am saying? Follow the heart, dear one. There is no wrong decision - how can there be? And that which is judged to be bad is usually that which is a result of fear. When you are to judge somebody for their behaviour - however that is - it is to know, indeed, that person is merely a reflection of who you are.

Q: So seeing a quality in somebody else is really seeing a part of yourself, is that what you mean?

P'taah: Indeed. Now, it is for you to understand this: That in all of your lifetimes, thousands and thousands - you have been everything - you have been that which is murdered and that which is the murderer. You have been the conquered and the conqueror. You have been the slave and the enslaver. You have been the rapist and the raped, you have been the child and the child molester. There is no thing that you have not been - you have been it all. So it is for you to remember this and to understand what pain and what agony and what incredible fear is behind actions that are not of love, not of compassion. And even in this lifetime, there are aspects of every one of you which you would certainly prefer that nobody else would know about. But you see, there is no judgment. All of you are created from the thought of God/Goddess, the All That Is. Every facet of who you are is an expression of that, else it would not be. *(A long pause).* Questions?

Q: (M) My question concerns manifesting. How is it that one's attempts at manifesting work only sometimes and at other times not?

P'taah: Now, this is a very good question. At this time there is much written and spoken of among your people in terms of manifesting. And you are told to do affirmations, to say 'this is what I desire, I desire to bring forth such and such'. Indeed, if you do it often enough it will come. Now, you may speak forth the words, however, dear one, the bottom line is that *you manifest by embracing thought with emotion.* If you are desiring much affluence of coin and each day on awaking you say 'I desire to bring forth much abundance of wondrous money' and if indeed it does not come, then it would behove you well to look at why. Now, we use money as an example, because it is indeed a great obsession with humanity. And we would ask you this: What do you really think about money? Now, we hear many things that people are saying about money and indeed our woman was speaking of this in the days past, and it was a very good exercise, because she wrote down what she thought and said about money. And she was amazed: It was: 'the bloody money...there is not ever enough of it...it is never there when you need it..you have to work for your money...money does not grow on trees...money is the root of all evil.' Does it sound familiar? And then you wonder, indeed, why it does not come when you are asking for it.

Now, we would also ask you to consider this: Money is an idea construct. In truth it is a symbol. It is energy. *It is a symbol of consciousness.* You really use it as an exchange, indeed? However, because of how you feel about it and what you think about it, it can be likened unto an entity. Do you understand? Now, if you are putting forth the belief that it is not there, that you really do not like it - another very good one is that money is not spiritual and if you are truly spiritual, then you are above money and you should not have it - so if you are pushing it away from you, how can you expect it to come to you? So what I am really saying about this, is to look at bottom line belief structures. *It is your beliefs which structure your whole reality.*

Let us look at relationships. Many of you who are not in a relationship - we would say a marriage or a love relationship - desire grandly to have a wonderful lover, but one does not come. So you are putting forth the desire: I desire to have a wonderful lover. But what do you really believe about love? The bottom line is, you believe that love hurts, that love is pain, because that is truly what you know. You believe that a relationship does not last. You have all of these beliefs about love and then you wonder why it does not come to you. Do you understand? So, when you are desiring something and it is not manifesting, take your paper and write down on one side of the paper what you want. On the other side of the paper, write what you believe and you will be very surprised, hm?

Thought embraced by emotion manifests your reality. Belief structure - that is all it is. Your beliefs. They are the building blocks of your reality. You believe that the floor that you are standing upon is solid - it is not. You believe that the wall is solid - that you may not walk through it. It is not solid, beloved. It is your belief and it is a common belief of humanity - solidity. You believe that your Earth is solid - it is not. There are grand civilizations living beneath your Earth. If your scientists would probe many miles into the Earth, they would find it is solid. But truly it is not - it is only the belief. The civilizations who live beneath your Earth are indeed of another space-time continuum. But they are indeed there. It is merely that you believe they are not. That which is your Earth is truly not solid, even beneath its own entity. Your whole reality is fluid. There is truly not very much difference between that which is air and that which is water or earth. Common belief makes it so. Which is very good, dear one - if it were not so, it would be very messy, hm?

Idea construct[2] - the beliefs held by humanity - also evolve and in this state of fluidity we will speak, for instance, of the physical entity of one who was upon your Earth named Yeshwa Ben Joseph, the Christed One. Now, about that which is your religion, we would say at this juncture that we are not really in approval of the religion upon

---

[2] The term 'idea construct' is perhaps grammatically not fully acceptable (along with a number of other special expressions of P'taah), but remains unaltered throughout this material. With this phrase P'taah refers to all those human realities (belief structures) which have little or nothing to do with truth or reality.

your plane, because it has been a structure for great enslavement, but that is on the side. The one who was the entity Yeshwa Ben Joseph really does not bear much relationship to that which has become the idea construct in these two thousand of your years. However, the idea construct is valid. It has created an entity, who in its own lifetime was nothing like compared to what has become the idea. It is no less powerful for all of that. There is nothing, no thing in your reality that you perceive, and indeed that you may think, which is not valid. Everything is valid and everything is divine expression. Even when you do not approve. This is discernment. You understand that indeed there are many modes of behaviour, of belief structures, which are perfectly valid, yet not for the greatest harmony for humanity or for the planet. However, *that which you invalidate you empower*. It is like fear. *That which you fear you draw to you*. Humanity believes that thought belongs in the head and has no reality outside of the head. But it is not so. Thought is energy wave. It has what you would call electromagnetic energy, drawing to itself whatever it is comprised of. The universe does not judge whether it is good or bad, it simply IS. As a thought goes forth into the universe so you co-create your reality. *If you are putting forth a thought embraced by the emotion that you desire such and such, the whole universe would re-arrange itself to create it*. In that fashion you draw to you the people who indeed are of like desire. *There is no such thing as an accident. There is no such thing as a coincidence. It is all created, co-created, indeed*.

So, there are many who would say: 'What of an accident, where a child walks in front of a vehicle and is killed?' 'And what of the parents of the child, and indeed, the one who is driving the vehicle and the family of that one; how is it that they would desire to create something so horrific for themselves?' 'How is it that a child will create a birthing in a country where it will die of starvation?' Indeed, *it is for the experience*. You see, there is no such thing as death in truth. It is a grand illusion. As we have said before, it is not one life, it is thousands upon thousands of lifetimes, and outside of this space-time continuum, they are concurrent. So you may say that if the lifetimes were drawn, as you would draw a line on a piece of paper, and you are looking down upon it, you can say the one looking down is called the soul energy, looking along the line at lifetime after lifetime. But in truth there is no past and there is no future, there is only now. Now, we are in understanding that it is very difficult for humanity to understand lifetimes as concurrent, as being simultaneous, as happening all at the same time. But you see, it would be very difficult for you if there were not, in this state of consciousness, time as you understand it. And, indeed, on other planets and galaxies there is also time, but it is a different time, and also there is knowledge, because those of the star people may travel through time as you understand it. So they may also see what you understand as your future; although do not ask us about your future, because we are not soothsayers and, indeed, you create your reality moment by moment. Now, I do know there are questions about this, so, please, ask.

Q: (F) Is it not so, that with every thought that I have, I create my future? So with any thought I have about you or the future, as I can choose my thoughts I choose my future?

P'taah: Indeed.

Q: Everybody does, so we have many different realities, and as every reality is valid, I should not be concerned about anybody else's reality. Is it my function as a child of God to create my own reality?

P'taah: Indeed, it is so. Now, we will say another thing about what you term to be the future, and we will talk very briefly about probable realities. At any time in your life, where you come to a choice point - now we say choice point meaning that you may do this or this or that - with each choice there is an emotive thought. Now we will give an example. A young man meets a beautiful girl and they decide that they would like to be married. Now, before the young man asks the beautiful girl, he thinks about his life and what it will mean to be married. On one hand he loves her very much, yet on the other hand he would like also to travel the world. He would like to be free to sing and dance with others and he knows that it will not be possible if he be married. He also, perhaps, would like to continue with his school and he knows, that this will not be possible. So, he weighs very carefully the desire of his heart, and perhaps, that which would be practical, hm? And so he decides to get married. And he does. However, he has already put in motion the other life, and because it has been embraced by emotion it continues. Valid reality. And the only reason that he does not consciously know of this, is that his focus is exquisitely fine-tuned. That is why we say to you that - truly - your life and your reality is fluid. It is not set in concrete. So you have also, you see, thousands upon thousands of probable realities, as does your Earth. So it is what you would call mind boggling, indeed. But each is valid.

Now, also there is much spoken about the change from what is called third to fourth dimension or density in this time to come - and there is much terror about the changes and what will happen during this change-over. Many are afraid that if they are more spiritually advanced than those they love, they will move forth into fourth density and those who are not so advanced will stay in third density or will die. This is human thinking in absolutes, thinking in concrete. It is fluid, there is more than one Earth - there is already fourth density Earth. There is already every dimension. It is not that if the people of the planet do not come up to scratch, so to speak, that the world will explode and it will be the end of everything. First we would say again that that is an illusion dear ones. You cannot die. There is no ending. There will be a time of transition, which you may say, in truth, hm, is like softly passing through veils of mist, where there is neither one thing nor the other. There need not be holocaust, there need not be great devastation. Certainly there will be gradual changes and some of these changes are coming very soon in your time. They have already started, indeed, and we would remind you, it is a feel of wonderment, these changes of your grand Goddess. Wonderment calls for joy. And when you are reading of those who die, know that it is choice, that there is no death. It is for the experience, that is all. There is no death - no end. Question, dear one?

Q: (M) Have we been, in a previous incarnation, that cat, that tree or that bunch of flowers?

P'taah: Indeed. Now, it is not to say that the soul energy - total soul energy - is the cat, the flower, the tree. But at soul level, the soul may choose to experience part of itself as anything it wishes. It can be a grand crystal, it can be on other dimensions of reality entirely. You see, you are not merely structured in your grandness only as a third dimensional human being. You are many things - whatever you choose at soul level. But we are not saying that when you come back, dear one, you will be a cat. That which is animal - and we are not speaking of Cetaceans[3], because that which is whale and dolphin is the same soul energy as that which is human - so that which is flora and fauna is of the same dimension which you may call second density. So, it is not that the whole of a soul energy would choose to become a tree or that which is second density or energy, but indeed, the soul may choose to experience itself anywhere, however it desires. You see, you are very powerful.

Q: (M) Can we influence our thought process on a soul level towards a certain desire then?

P'taah: Are you speaking of incarnating as second density?

Q: Rather fourth density.

P'taah: It is not so much that your third density beingness will influence so much that which is soul, but rather the soul will influence you in this understanding of third density being. But certainly you may ask forth of the soul, the God/Goddess of your being, that you would desire such and such. There are times when that is alright. But you know, in truth that which is the greater of you knows what is in your best interest.

Q: So, by aligning with the God within we create our best possible path in any case?

P'taah: Dear one, you always create the best path in terms of the lessons, that are to be learned. Do you understand? But of course, as you become more and more aligned with the divine thread, if you like, then you are truly always in understanding of what is most harmonious. Indeed.

---

[3] Cetaceans: (Latin: cetus) a whale. Mammals of fishlike form, including whales and dolphins.

Q: Is it possible to physically die and remain conscious about it and come back?

P'taah: But it happens very often. There have been energies who have been clinically dead and did, indeed, become what is called resuscitated.

Q: I am rather thinking along the line of the many people who have fear of dying. I remember my father who passed away earlier this year and who did not want to go. He had fear of leaving. So if someone dies - I would volunteer - and could come back to show others how it is done, then fear would be greatly reduced. So one could give help to the aged and distressed.

P'taah: Dear one, even if you were to die and to return - and you know it has been done - it has not helped people in their fear of death. That which is your Yeshwa Ben Joseph indeed died and returned, hm? It did not help many people in understanding, that truly there is no such thing as death, that it is illusion. So, beloved, even if you were to do it, you would not be helping people, because each must come into that understanding for themselves, because each is creating their own reality. Now, there are many who go through the physical translation[4], who believe implicitly that there is heaven and there is hell. And they have a picture in their mind of what is heaven and what is hell and when they make their translation, that is exactly what they find.

Q: Do you mean transition?

P'taah: Indeed. Then they will come into the understanding that it need not be like this. There are some who believe that nothing occurs after death - well, they are very surprised.

Q: So, what happens really after one dies?

P'taah: The first thing is that you take your consciousness with you. Indeed, that which is your energy - you take it with you. Many people are not truly understanding that they are physically no longer, because they still feel as if they possess a body, and for those of you who have had out-of-body experiences, that is where their consciousness is leaving the body and they turn and see the body

---

[4] One of P'taah's terms for physical death.

lying on their bed, they say: 'How can this be, because it feels still as if I have my body'. Hm? So it is an extraordinary sensation for many people, but your consciousness goes with you. Now, that does change, however, it is very difficult for me to speak to you about what occurs truly, because it is like explaining to a foetus in the womb what it is like after birth.

Many of these concepts become very difficult, because we are really speaking to what are limited boxes of your belief structure, of your conscious knowing. And so, what we are doing is to attempt little by little to expand that consciousness, so that you will come into a greater knowing - that you may tap into the greater knowing of who you are. And dear ones, also know this: As you ask your questions, and you may ask anything - nothing is considered too trivial - but as you come forth with your question, already in truth you know the answer. Now, beloved ones, we are in understanding that you have many questions, however we feel that this is sufficient unto the time. There has been much information this evening and we do not want to blow the circuit - overload the computer, hm?

We would say it is always joy to be with you. Humanity indeed is beloved of my heart and we are in anticipation of our next being with you. So, you may be very busy with your papers preparing many more questions and we will try to give you the answers to the best of our ability.

(P'taah turns to a silently weeping lady). It is alright, beloved one. Not to be taken too seriously - to know that indeed it is you who has chosen this wondrous experience called life, and that which is the separation makes this communion that much more dear.

Dear ones, go forth in love and laughter. We would say play more, be silly. I am not jesting. Well, of course I am jesting, but also I am not. Our thanks.

Chapter 2

SECOND TRANSMISSION.

Date: 6th of September, 1991.

P'taah: Dear ones, good evening. How are you all this evening?

Audience: Very well, thank you.

P'taah: Well indeed it is joy for me to be with you. *(P'taah briefly addresses the person in charge of the sound equipment and the transcription of the tapes).* Your device is in order? Very good. We are indeed, dear one, hoping that you can make some sense of these words. You know, that which we spoke forth thus far, has been very much that the energy is one to one, because indeed we communicate in more ways than only the senses which you are in understanding of. We speak [as well] to other dimensions of who you are, and so it is that you have an understanding that comes within you - as we communicate in this fashion - that is very different from reading the word. So it is up to you to make sense of it, dear one, alright?

Welcome, dear ones. Now, the foundation blocks of your reality are belief structures. They are the bricks, if you like, with which you build your day to day reality and, indeed, your known universe. In many respects, we are addressing what is called basics and you have heard the words before many times. However, many of the basic truths that we speak are not really set in concrete emotionally, although you may indeed have a wondrous intellectual understanding. The building blocks of your reality: belief, hm? The building blocks of who you are is your belief about yourself and those blocks are set in place very soon after your birthing. You are taught what to believe about who you are. Now, it is unfortunate that you are shown who you are in what is called negative aspects.

Your role models of the past in your culture - and, indeed, it is different for different cultures - are so wondrous, that you feel that you may never ever match up to perfection. Now, we are speaking specifically here of the basic religion of your time. Now, religion has been intellectually displaced for millions of people in your culture. However, dear ones, emotionally it is with you. It is also with you in what is called a collective consciousness and we are speaking here of Yeshwa Ben Joseph, the Christed One. You are taught from an early age - whether or not you are literally taught within your home - that this One died because you are so bad, died to save you, hm? It is not so, you never needed saving, and that which was the Christed One certainly did not die for you. Rather, as an experience for his Self, and if you like, as a demonstration. The purpose of the demonstration was to show humanity that there is no such thing as death. Well, you know how it has all been twisted around and so, for all of you from your time of birthing, you grow up in an understanding of guilt and sin, and thus judgment is born for you. Now, we are speaking of this at this time because it is very important that you understand these beliefs, which you carry forth from your childhood, from your experiences as a child; even if they were not, as one would say, traumatic, they are still the lens that you view your world through as you are growing and coming into adulthood.

All of you believe that you are not worthy, that you are not worthy of love, that you are not worthy of all the wondrousness that your universe may provide for you - that you are not worthy of the love of God, that many of humanity are not even worthy to draw breath. It is very interesting to observe the diseasement of those upon your plane, who truly find it difficult to draw breath, because they in truth believe that they are not worthy to live. You believe that you end where your skin ends. It is not so. You believe, and we have said this before, that your Earth is round and solid. It is not so. You believe that thought is something which stays in your head, that it does not concern the rest of your universe. It is not so. It is a very interesting exercise for you to learn what you believe, to discover what you believe about Self, what you believe about other people - what you believe about your planet - about your flora and fauna - what you believe about your universe. You know, when you start to really look at what you believe, you will be astounded at the belief that you carry, which intellectually you know to be nonsense. You believe that you are not truly part of God/Goddess, the All That Is, and we know the yearning within each breast to become whole. It is tangible - the yearning. And you know, it is also true that in another sense you are whole. That which is the separation of Self from SELF, that is the most bitter pill for you, for all, all of humanity. All people suffer the grief of separation, the knowing within that you are not whole and the desperation to be whole.

As you believe that you are not worthy, that you are not clever enough, that you are not beautiful enough, that you are *not enough,* so you are keeping yourself in separation. We have said to you that everything outside of yourself is a mirror for you, a grand reflection, and we have said to you, when you look to another and judge harshly, you are merely looking at Self, seeing the reflection of who you are. But you know, dear ones, we would say to you: Go into your garden, look at your sky, look at your trees, look at the Moon, the sunrise and the sunset, and in the beauty of that moment, know that it is also the reflection of who you are. You are wondrously beautiful, each and every one of you and you do not know that you are. *(At this point P'taah's voice has become indescribably gentle).* — You do not know — and we would have you all be in awe of your own beauty. To learn how it is to have a wondrous love affair with who you are.— How do you love who you are? *(A long pause)* — Indeed, how do you love who you are?

Look into the mirror and know that you are a unique gem, a facet of the jewel which is humanity, but all that you really know is pain and anguish and desolation - well, this is the time for the change, indeed. Humanity and indeed the universe, operates out of what is called masculine energy, that is the doing. Humans *doing* - and now is the time to be human *being*. Masculine energy is the striving, the organization, the thrust. Feminine energy is that which is nurturing, allowing - that which is giving of itself. Now, it does not matter what gender you be. *The soul has no gender.* Humanity and that which is woman have come from masculine energy, the striving to do, to survive. The women of your planet have been very joyous about that which is called woman movement - New Age - power to women. These women have been very happy. They say: "Now we will show

25

them - we have got you this time, you bastards.' Hm? You know it very well. But you see, the truth is that the women have come from masculine energy, because they had to survive and because in your culture feminine energy has been despised. Man to man. If a man were to cry, to be gentle, nurturing, to be allowing - it has been simply unforgivable. Well, all of you here have seen the changes coming forth, and it will be more and more. It is not to decry that which is masculine energy, it is to have balance - that man and woman may have the balance of masculine and feminine energy. It can be likened to the two wings of a dove, dear ones. The dove will not fly on one wing. Balance, hm? To be allowing of who you are. *To be allowing of all people to be who they are.* It means enormous changes for the education of your children. And very soon in your time it will be, that the children will not go to school. Indeed, it will be a very different type of learning experience.

When the balance is occurring, so it will be reflected upon your planet. In the meantime we may say this: There will be many things occurring, on a planetary scale, which will be most discordant. We have said before, that - as the changes are coming forth - many people will be in fear, in trepidation. And dear ones, you have chosen to be here, to witness the changes; to facilitate that the changes may be harmonious - that you may understand that *there is truly, truly no such thing as death.* And as you believe that there is no such thing as death and as you believe that you may change all things in your physical world - and as you come into the understanding that every entity upon the plane has power - how do you think it will be for you? When you know that you may change things, change your life, then you will not be in fear and you will teach others that it is not necessary to be in fear; that the changes coming forth are what you would call of right order. That which is cataclysm is called 'opportunity' and that which may seem to be dire circumstance, indeed, may be cause for joy and for celebration, because it is all bringing about change for the new order.

It is in this fashion that you may come into an understanding of how important it is to know what you believe and what you believe about yourself. *To come into the knowing that everything outside of yourself is a reflection.* As you become more and more integrated and are coming more and more into an expanded consciousness, so you will be a grand reflector for all about you. *You create your own reality absolutely.* There is no accident, there is no coincidence - it is you, dear ones, it is you. And you have the power to create a wondrous, wondrous paradise.

Now, in these coming times, when we speak together, we will be covering piece by piece of how you are in your life and how it is that you may create a change within Self; and how it is that you may embrace that which is anguish and pain - how you may change the dis-easement of the body and how it is that you may become integrated, to become whole, to become more and more and more of who you really are. Now, questions, dear ones.

Q: (M) When you speak of the new order, do you speak of the same as the politicians, when they speak of the New World Order? I believe you do not, but I would like to have confirmation.

P'taah: When we are speaking of a new order, perhaps you think that this is not the correct terminology. A new order of humanity. In fact we would almost say a new species of people, that is what has been heralded as the change from third to fourth density, or dimension, of reality. Now, we would say this: We do not really like to speak in these terms, *because what we are speaking of regarding this change is truly beyond the comprehension of who you are now.* There are many of you who have had glimpses of what it may be like, but as we had said before, beloved ones: *To describe how it may be, may be likened unto describing to a foetus within the womb what life is like after birth.* However, you know that you really know how it is, because you are already there, beyond your conscious knowing at this time. However, that which is the new order is that you will consciously know, that you will become in your own conscious understanding the beings of light - that you may come into grand technology, *that you may know indeed that all, all is One.* That there is no separation within yourself and from person to person. That you are all from the Source, indeed part of the Source - divine expression

- that there is no separation. You will understand how it is that you create physical matter, that you are the thought of God and in that understanding you will come to know manipulation of matter; that you may travel within galaxies and that you may choose *how* you may be and *where*. As parts of your soul energy are hither and yon, you will know that your communication will be telepathic, that you will be able to read the light energy of every atom and molecule. That is the new order, beloved.

Q: What you say makes one's heart jubilate.

P'taah: Indeed. *(P'taah walks up to the questioner and gently touches his forehead).* You see, dear one, you have had the experience of how it is to be One with all things. And sometimes it is more difficult when you have had these experiences, because always, always you are in grief and waiting. *(There is a long pause as P'taah is kissing the forehead of the gentleman.)*

(P'taah turns his attention now back to the audience). Questions?

Q: (M) So the process is happening already and it is just a question of allowing things to happen and letting them unfold?

P'taah: Indeed. That is precisely how it is. *Nothing to do, dear one, simply to be.* Do you know that there is no tomorrow? There is only now and it is for each human to live in the now moment, in the fullness of the moment - in the joy of the moment and even if it not be joy, dear one, to certainly be in the fullness of however it is. And it could be that it is grief and pain, *but it is to be embraced to the full*, because it is only in this fashion that you build your tomorrow.

You see, your concept of time is indeed localized - local. Outside of this space-time continuum there is a different time and that time is only now and that can be any time. We would have you remember that there is no tomorrow and as you are living in your past and are worrying about the future you are not living *now*. It is by living now that you may allow the tomorrows to be in the fullness of their own beauty. Allow, hm? Grand feminine energy - to allow. It is extremely expansive. You know, you create every next moment from this one and if you are worrying about how it will manifest and you say: 'I want this and this and this and I want it to occur in this and that fashion and by this time or that time.' - what do think will occur? What occurs is that you immediately shut down millions of possibilities. The great art of manifesting is to say: 'I desire from the God/Goddess of my being that and that.' Put it forth to the universe and *know that it is already occurring.* It may take a little time to manifest, but that is only because you are truly not knowing *how* it occurs. But when you cast forth the desire embraced by the emotion, it is to know that once you have cast it forth it is already in occurrence. It is very simple. When you become dogmatic and determine this and this, you have immediately changed the possibilities, because there are always millions of variables. Questions?

Q: (F) Relating to the raising of consciousness, how important is the body or where does the body come in?

P'taah: Well the body is indeed a wondrous vehicle. A temple to house the soul, hm? *Your body was designed to last hundreds and hundreds of years.* You believe that it will not and it does not. Your body is a vehicle to experience this dimension of reality. Your body has its own integrity, it has its own cellular memory. It has a wonderful resonance with your planet, indeed, there is no fabric of your planet which is not of your body. It belongs here. Now, we noted that the body is very often treated with great contempt. There are many grand spiritual people who believe that their body is not important and indeed, they would do anything to get rid of it. They feel that the body is the only thing stopping them from reaching paradise. These people do not understand that they have chosen their bodies as a grand expression, which is also a divine expression. Without the body there would be many, many joys lost to you. The body has been going through many changes and there are the changes which occur with that which is pollution - ill-treatment. However, the body is indeed a wondrous instrument, it is a wondrous entity; resilient, creative, wildly creative. And when we are speaking of integrity we are speaking at cellular level. *The body is Spirit made manifest.* Do you understand? It is a call for grand celebrations. It would be a very boring third density if you would not have your body. It is interesting for you to know what you believe about your bodies, and the connections between your bodies and what you believe about yourself.

Humanity is made up of sexual beings. You have gender and we have noted there is extraordinary turmoil within humanity about sexuality. There are many beings who believe that their bodies are dirty. We have not noticed that their bodies are dirty, unless you have been in your garden, hm? Your body speaks to you all the time. Your bodies are like a barometer. That which is diseasement of the body is merely indicating a dis-easement of the Self. It is interesting to look at the parts of the body that become dis-eased. As you become more and more into the knowing of who you are, so you will become more and more in tune with what your body is telling you how you are. Now, there are many people who will say that this and that is not good for the body; the type of food that you eat or what you may or may not drink. We would say to you that your bodies are very hard to kill off - almost impossible *if you are in joy and if you are truly in harmony you may indeed eat poison and you would transmute the poison automatically.* We will speak more about the process of transmutation after this evening.

Dear ones, we will take a break and you may prepare yourself for more questions, hm?

Q: (M) Just before you leave, one question: You are a grand teacher to us. Do you get anything out of this?

P'taah: *(mocking outrage)* Beloved!! It is never what is called a one way street, you know? But of course! You are wondrous teachers for me. It is a great joy for me to be here with you. It is a grand learning experience for us - and we love you so much, how could we not be here with you? Indeed.

(P'taah takes leave to afford the audience a short respite.)

(After the break:)

P'taah: Well, dear one?

Q: (F) Do you ever get upset, that even though you say it and state it, we still forget, that we hear the words and we miss the depth

and the reality of what you are teaching? Do you ever get bored with us? I know I get fed up with my forgetfulness.

P'taah: Beloved woman, how could I be bored? Each and every one of you is striving to come into the knowing and of course, you forget. What is occurring at this time is that you are undoing that which has been since eons in the making. *That which we speak of is diametrically opposed to everything that you have been taught lifetime after lifetime after lifetime.* What we speak of is indeed diametrically opposed to the morphogenic[1] consciousness of humanity. *It is a quantum leap in consciousness that you are heading for.* There is no boredom in this. But, indeed, it is so that in the understanding of your yearning and of the pain and the anguish and your broken hearts, that we indeed are also broken hearted when we say the words and you truly do not hear. But we are also very patient and we are also in the knowing and understanding of who you truly are and we understand where you are going, because in another sense you are already there and this we know. — You have another question, beloved?

Q: It is a little more personal. I have been in fear and some anxiety about survival this week. I have been striving and trying to push and while I understand that this is also who I am, I have a preference to surrender to that wonderful energy, which I know is there. That energy is so expansive but I forget. I cut myself off from the beauty of this place when I am in that striving place.

P'taah: Therein lies the lesson, beloved, hm? Now, how fortunate are you, beloved one? That at least you know. For all of you, who are into the striving, there comes the moment where you will say: 'Indeed, what the hell am I doing - where is the love and enjoyment in my day-to-day life? Why do I get so caught up?' And then, indeed, it is to take the quiet moment to remember. Sometimes the only quiet moment that you have is when you are in your bed - before you are sleeping. It is then that you may, before you drift off into your dreamtime, think about your day and be in understanding of what

---

1 Morphogenic resonance or consciousness: Collective or mass-consciousness or global thoughtform.

you have let slip from your consciousness. Merely that. But you know, it is not to berate yourself and it is not to feel guilty. It is to embrace, always, who you are - to know that it is alright and to know that you may create more harmony for Self, if you will just take the moments. It does not need to be long. To take the moment for Self - for integration. To look about you, to look at the sky, to listen to the birds. *(At this point, somewhere out in the night, a rooster pierces the stillness of the evening with his crow and P'taah remarks:)* Even those ones. *(The host jestingly disagrees, due to some experience with 'those ones', and remarks:)*

Q: Also at three o'clock at night?

P'taah: Indeed. Indeed. And when you are awake at three o'clock in the night and you hear the cock crowing, it is to rejoice - he is not so concerned with time.

Q: Very true.

P'taah: Those of you in this place are very fortunate. You live in very beautiful surroundings and you have chosen it for its beauty, so it behoves you indeed, to enjoy it. That which is nature will teach you more than any of your books, any of your films, any of your gurus and certainly more than I may teach you. — Questions?

Q: (F) P'taah, concerning dreamtimes, can you tell me why it is that sometimes our dreams tell us exactly what is going to happen the next day - why can it not be like that all the time?

P'taah: But you see, beloved - that which is sleep allows you to experience your multi-dimensionality not only for the lessons, but also for the grand adventures. As you are sleeping you are travelling, you are meeting people you are not in physical contact with, both on this planet and plane and other planets and planes. You are very busy in your dreamtime and indeed, if you did not have it, you would not exist, because that is the true reality in another way. It is real, it is valid, and when you are in your other states of consciousness it is this that is the dream.

Q: How can we retain that consciousness when we awaken? Can we practice a technique to achieve this?

P'taah: It is not necessary, beloved. It is not necessary. What you are striving for is to come into the understanding of multi-dimensionality within consciousness and this, indeed, is occurring. But certainly there are things that you may do to enhance the lack of separation. One of the things is that you do not need long blocks of sleep. If you were to sleep no more than six hours of your night and then indeed, if you were to break that time up so that you were to sleep in three hour intervals and then were using the hours of your early morning after midnight before four a.m., then the separation of the realities becomes less. You may use the time of your early mornings to come into altered states of consciousness, to have adventures, to consciously leave the body - you call it 'out of body experiences'. This becomes very easy in these early hours of the morning. You may put yourself simply into relaxation and ask forth, before you sleep, that you only wish to sleep three hours and then you wish to play. And then you may experience these wonderful states of consciousness, although it would be beneficial if you would write down your dreams. And if you are waking within that split timing of sleep periods you will find that your dreams are very vivid, and if you write them down, then [do this] certainly for no less than twenty one of your days. Many of the people I know are keeping dream books and always write their dreams down, but after a time you would find the dreams to be in a pattern. That pattern will be like your unconscious mind speaking to the conscious mind. It is a grand adventure and we will say, certainly worth the effort.

Q: (M) Tying in with this subject, I desire clarification regarding a recurring 'out of body' experience, during which I rise to various heights in a physical sense. Last night, for instance, I rose as high as the outer atmosphere. What is actually happening?

P'taah: Indeed. Now, there are times when you are travelling within this plane of reality, and that is when you do not feel that you are rising very far. If you are being conscious of this, then you may project your consciousness to any part of this globe. You may drop in on friends. There are other times when you go beyond this plane of reality. In your terms it would be beyond Earth. Now, when you feel the experience starting, you may indeed ask that you wish to have conscious knowledge of what is occurring and you may do this whether you are asleep and dreaming or whether you are awake and experience an out-of-body experience. Because very often, even when you are having an out-of-body experience - that is consciously, where you may look back and see your body - if you are travelling very often outside this dimension of reality you will not bring back the conscious knowledge. Now, that is not to say that it is not affecting your day to day reality. It is just that you have no conscious knowledge of this. But when you are dreaming you can direct it and you may be able to say in your dream: 'I wish to bring forth the consciousness, I wish to know about the adventures.' You can do this - you can manipulate your dreamstate. You may be conscious of dreaming within a dream.

Q: *A side effect of this experience is a sensation of healing, a feeling of extreme well-being. How does this relate to it?*

P'taah: But it is so. When you are travelling forth and when you are in communion with your brothers and sisters from other dimensions of reality and you are tapping into great wisdom and great joy, what do you think occurs, dear one - what do you think is the benefit? It is creating a healing for you. *Ask and it shall be made manifest unto you.* Do not try. Ask and *be allowing without expectations.* As we have said before: expectations close down the probabilities, narrow the field of manifestation. And that occurs on every plane. Alright?

Q: *And is there any risk involved of not returning to the body?*

P'taah: There is no risk. It is a very valid question. There are many people who are terrified that they will not find their way back. We would say this to you, beloved ones: *You exist in a safe universe. Nothing can harm you. You cannot die.* You have great integrity of soul. You are sovereign beings. Nothing can harm you. Wherever you travel in the multiverses you are safe. You are safe in the integrity of the God/Goddess of your being.

Q: *(F) As a follow-on of Julia's and Gita's question a while ago: When we become physically ill, what is it that causes us, perhaps our thoughts, to initially get off the track, so to speak?*

34

P'taah: Dear one, it is not thought. It is called emotional dis-easement. *Your physical body is an expression of your emotional well-being or dis-easement.* Now, it is also a reflection of your belief structures, your beliefs about yourself, your beliefs about your environment. If you believe that you do not live in a safe universe, then everything in the universe will indeed re-arrange itself to accommodate that belief. Do you understand? If you believe that you are a victim - you will be a victim. What humans are not understanding really, is that as they emotionally reflect pain, so the body will reflect that pain. Now, we are getting into areas which I would prefer to address a little later. However, we will cover it briefly at this time. But this information would really best be categorized in a future session.

As you are birthed and as you are living with invalidation and as you believe about yourself that you are not enough and that you are unworthy, so life reflects to you that belief. It creates a situation, where you are in judgment of yourself and in judgment of a situation and your terror of the pain creates a wall of invulnerability. The terror of being hurt. We are using this as an example, dear one. And as you are invulnerable, so you create separation, the fear of grief and the fear of grief being inflicted by a person outside of Self. You create a wall of invulnerability, so that you do not have to feel. But, of course, it does not work. So the fears and the feelings are locked away, the emotion is kept in a bottle with a very tight cork, but it cannot be contained, because that which you are is thought embraced by emotion. *Where emotion is not given access to expression, the body will create a dis-easement to tell you to do something.* It becomes like a spiral, wherein people decline. We would take, for instance, one of the great diseasements of your time, that which is called AIDS. You know this one, hm? Look at the people who are creating this for themselves. You know, nobody *'catches' a disease.* It does not work like this. Humans have constructed an idea that as they are in their fragile, little bodies racing around their planet, they are vulnerable to these nasty little viruses which invade the body. That is not so - you are not victims. You create. The consciousness of human beings will tap into a situation and create that which is expressing the emotion. Male homosexuals have a great stigma in your society, which is quite extraordinary, because in truth human beings are most naturally called bisexual. We understand that this is not a happy thought for many people, however, it is emotional truth for you. The people who are injecting chemicals into their bodies are also very vulnerable to being invaded by this nasty virus. How is it for them? We are speaking about guilt, we are speaking about lack of acceptance, love and support - we are speaking of a sense of unworthiness - we are speaking of people who are involved in great trauma in their childhood. We are speaking of people without hope, without joy. Then along comes *this dis-easement*, which, we may say, *was created by one of the governments of this planet for the purpose of annihilation and distributed by an agency of one of your governments* and look at what it has wrought.

There is also what is called childhood dis-easement and everybody is rushing around inoculating against these diseasements. There is also the field of medicine itself. We find it most extraordinary that women are encouraged to examine their bodies for diseases. We do not think it is very sensible. To look for a dis-ease? Do you see what is happening when you understand that you create your own reality absolutely? That your thought embraced by emotion creates matter, creates physical reality? Then you may tell me how sensible it is, that a woman each month will examine her body for a dis-ease. It is not very sensible. To inoculate against a disease which is rampant, what else is it but encouragement. But you see, it is in the morphogenic reality, within the collective consciousness. Now, we are not saying: 'Do not use medicine.' We are not saying this, because it is the belief of humanity that it works. But we are saying, that as you become more in tune and more and more at ease with your reality, more in understanding of how you create your reality, you will find there is less and less need for medicine. But it is valid, let there be no mistake about this, dear ones. There is no judgment about this. We are merely pointing out to you how it is - the isness of it and the isness is that it is in existence, therefore it is valid and it is a divine expression, else it would not be here. When you are finding your body in dis-easement, it is to ask why, what is the emotion behind the disease of the body. Then it is to embrace that emotion and indeed the physical diseasement. Know that it is only there for a lesson. And by the embrace you will create the change. We will speak more of this soon. Is that in answer to your question, beloved?

Q: Yes, thank you.

Q: (M) P'taah, are there any other planets, whose inhabitants have chosen separation as much as we have here on Earth?

P'taah: There are.

Q: Could you elaborate, please?

P'taah: It is not necessary, but of course there are other people, many of whom are technologically advanced on their planet, but who are not very advanced, as you would say, spiritually. And where people are not spiritually knowing, there is separation. We will say this also, and it is very important for your people of this planet: There is at this time great interest in technological advancement and we will probably speak more of this later also. However, we would like to say at this time: As technology advances, so there will be great changes. Much of technology - we would say crystal technology, that is using machinery without using moving parts, and we are speaking specifically at this time of transportation, but much, much more - the power, the fuel if you like, is thought. Where there is crystal technology there is enormous magnification. There are two forces, which are creator forces, upon the planet, in fact in all of reality: One is fear and one is love. *Where you have technological advancement without spirituality we are talking of fear.* Where fear is within the thought of those in power of the technology, the fear is magnified millions of times and thus will create devastation and desolation. Where the technology is powered by thought and embraced by love, it is the love which is magnified. Until humanity is coming into the understanding of the importance of thought, the importance of understanding that you are divine expressions, the knowing of how your universe works, this technology will be the devastation of your planet.

However, beloved ones, it is not to be feared, because you have chosen. We are here really to help you along to bring this wondrous dream into fruition, so that you may come into the grand technology, that you may come into the knowing, the understanding, into balance, so that humanity may indeed join with what is called the Grand Federation of souls who are waiting for you. *It is time for your planet to take its rightful place.*

Q: (F) At this time it saddens me to witness the enslavers, be they of other worlds or this one. I know we should love and embrace them. I can live with nature, but find it very difficult when I look at what is happening in this world. It saddens my heart...

(The lady asking the question is, for the moment, overcome by her emotion).

P'taah: Indeed, beloved. That is certainly becoming more and more so. We will say this to you: As humanity is moving into an understanding, so will the resistance grow of those who have no desire for this to occur. Let there be no mistake about this. There are many who will resist and resist and create as many problems as they possibly can in their terror, in their fear. And you are quite right, beloved. The only way that you will create the change is with the embracement, and to know, truly, that whatever is not an expression of love is an expression of fear. Those who are lusting after power, indeed, are in terror that they are powerless. And those who are wanting the riches of the world are in fear and terror that they have no riches and they do not understand the richness of Spirit. And those who indeed would enslave you, are those who are in terror of being enslaved. So it is to embrace it all and in that embracement you will create the change.

Q: So, in other words, if I am in fear of them enslaving me then obviously, in a way, I am trying to enslave them.

P'taah: Indeed, if you are in fear of being enslaved, beloved one, that is exactly what you will bring forth. *There is no judgment in the universe.* It does not matter whether you are putting forth your thoughts embraced by fear or embraced by love. The universe does not differentiate. You will draw it to you.

Q: That means it is alright what they are doing and once I know that, I don't need to be sad.

P'taah: Exactly. When you understand what this creates for you, then you know it is called overview. It is taking one step back to review the situation - to know indeed that all of the people of the planet are struggling for whatever it is that they are in desire of and whether or not it involves enslavement or otherwise. As you are putting forth the thought, you are drawing to you exactly what the thought is in the emotion of it. What you are is vibrational frequency, that is *all* you are. Well, beloved, that is all you are, but indeed, that is the grandest thing of all. You are vibrational frequency, thought is vibrational frequency. Thought embraced by the emotion. As you understand that you live in a safe universe, that you are sovereign, that the only enslaver, in truth, can be yourself and as you understand that you draw forth to you that which you believe, that is how you co-create your existence. And that is how the people may learn one to another, because *when you understand that you live in a safe universe, that you create your own reality absolutely and when you understand that the greatest power in the multiverses is love and that you are separate from no one and no thing and that everything that you behold and indeed everything in the known and unknown realities is an expression of the All That Is, so you will be a light unto the world and create the change that you all desire. And as you are desiring the joy and as you are desiring peace and tranquillity for yourself and as you are desiring to create paradise on your beautiful planet, so you make it manifest. And it is to know within your hearts truly, that nothing really can be destroyed. What you are is sacred, what your planet is, is sacred, what you are is beauteous indeed - hm - and what you are, indeed, brings joy to my heart.* It is to accept who you are, beloved ones, it is to accept and acknowledge every facet of who you are as divine expression. You see, everything comes back to you and there is nothing to do. It is simply *to be* in the knowing that there is nothing that you can do which is wrong, that there is nothing that you can be which is not divine expression. And as you come more and more into the acknowledgement and acceptance of who you are, so you may acknowledge and accept all. This is called sovereignty and this is called freedom. And it is called going forth to the light, it is called coming home to who you really are.

So, indeed, dear ones, sufficient unto the time. I desire that you would go forth in love and joy, not to do - simply to be. And when you look into the mirror in the morn, look into your eyes and say: Indeed beloved, you are expression of divinity.

And so you are indeed. Good evening to you.

I love you.

Chapter 3

THIRD TRANSMISSION.
Date: 11th of September, 1991.

P'taah: Good evening. *(P'taah's dynamic manner of greeting brings everyone present to fullest attention.)*

Well, dear ones - joy to be with you again, and for those of you for whom this is a new experience, you are welcome - indeed, you are well come, because the timing is of ripeness.

Now, this evening we are to address relationships to begin with. The relationship of Self with SELF and, indeed, the relationships with those beings who you draw into your life - for whatever reasons. We have spoken before that you *create your own reality absolutely*, that, indeed, *there is no such thing as an accident, there is no such thing as coincidence within your life.* We have spoken to you about your personal reality and the collective reality, both of which are based upon your belief structures. We have said very often to you, that as you believe about Self, so it is that you draw unto yourself exactly what it is that you do believe in.

Let us speak for one moment about love, that which you all desire so poignantly and which, for most of you, escapes you so readily. As you believe that each and everyone of you is not worthy of receiving love - and we will address this more - so it is that you create this in your life. *If you, as humans, indeed loved SELF, if you were in love with who you are, if you were in total acknowledgement and acceptance in the knowing that you are divine expression - every relationship would indeed be a loving relationship.* That it is not, is a direct reflection of *how you view yourself* and how you view yourself reflects everything outside of Self.

Most humans indeed have a very limited scope in their understanding of what love is. When a child is born, then in the beginning there is great love and tenderness - not always, but most often - for the new babe. The apparent helplessness evokes wondrous

41

tenderness and compassion. As the babe is growing and becomes a person of its own will, so are the social mores imposed upon the child, wherein the parent expects the child to be what the parent wants it to be, without regard to who and what the child wants to be. Very often this is felt by the child as lack of love; and it does not matter what wondrous parent they be. As the child grows up with the understanding of its lack of Self-worth and a lack of true communication with its parents, so indeed are the misunderstandings perpetrated. That which is a love affair between man and woman, or indeed between woman and woman and man and man - it truly does not make any difference, because in truth we are speaking of the love of the heart and however that is translated is alright - also brings with it the limitations of each person. But we would ask you to think of it this way: As you believe about yourself and as you believe about love and as you believe about other people, so the thought is broadcast through the universe, and thought - which is your most powerful tool and much more - has its own power, its own, what we would call, electromagnetic energy and draws to itself that which it puts forth. And so, as we have said before, if you believe that love is pain, if you believe that love affairs do not ever last, if you believe truly that you are not worth loving, then of course this is exactly what you draw towards you. It is somebody and some situation which reflects to you what you believe. It is very simple.

Now, there is also this to consider: Love placed outside of yourself - and we speak of what comes in humans closest to that which is unconditional love - is to show you how it may be to love yourself; it is a lesson, it is a reflection, because: *until it is that you truly can love who you are, that you truly can accept and acknowledge every facet of who you are, you do not know what love is.* As you come into the acceptance and acknowledgement of who you truly are, so it is reflected upon the whole plane, so it is broadcast into the universe - this wondrous love affair of Self with SELF - in that measure it creates the closure of separation of Self from SELF. So it is that, in the allowance of *being*, you are creating a balance of masculine/feminine energy within Self. The balance is reflected out there in your universe. So you see, beloved ones, whatever you may read, whatever you may search for, whosoever you may listen to, whatever wondrous teachers and gurus you follow, always, always, always it comes back to you.

In this period of time, where we have been communicating with the humanity of recent time, there is always great curiosity about the *'out there'*, hm? About beings from other places, whether it be on other star systems or whether it be the peoples of your inner Earth. Whether it be of grand technologies beyond the wildest human imagination of this time - always I will say to you that until you know who you are, you may truly not have the knowledge of that which is beyond the consciousness of Man at this time. In a way, dear ones, it is called responsibility upon your shoulders, because as each and every one of you comes into the greater understanding, so you are contributing to the greater knowing of all people.

(P'taah's following point refers to a principle called the Law of Resonance and Affinity, a certain aspect of which is - perhaps - better known as 'The Hundredth Monkey Syndrome'). The knowledge of humanity is within what is called morphogenic resonance, of collective consciousness. This resonance is changed when a certain energy - some say it is a number of people, but in truth it is more than merely numbers, it is more the vibration - when the vibrations of knowing come to a certain pitch so to speak, that a quantum leap is initiated, so that the whole of the consciousness of humanity leaps forward in knowing.

This leap does not happen by intellectual knowledge. Dear ones, we are not decrying the intellect, of course not, but we are saying to you that the intellectual knowledge cannot occur without the spiritual knowing within the breast of each and everyone of you. What you have all known, thus far, is imbalance. What you know in your culture now, is intellect which scorns emotion, that which is feeling. But you see, what you are is a vibrational frequency *empowered* by feeling. What you are is the thought of the All That Is, empowered by feeling. Now, in your culture of recent time, feeling, emotion, and that which is imagination has been looked upon with great scorn. Your imagination is the greatest tool that you have. It is the greatest creative impetus, and what you are *is* creative. Every atom, every molecule comprising your physical embodiment, is creative, powerful - of great integrity, else you would not be here. Your imagination can carry you beyond time and space. Feeling carries you also beyond time and space. That which is feminine energy is indeed what brings you once more into the understanding of feeling. To allow - feeling - to be intuitive - what does it mean to be intuitive? Many humans have no idea and indeed, we have heard very often great scoffing in the life here, when that which be woman says: 'I wish to do such and such because I have a feeling', hm? It is 'pooh poohed', it is not logical, it is not of the intellect, it is nonsense - it is woman stuff, eh? Well, we are very happy to note that this is occurring less and less. Humanity is coming more and more into an understanding that intuition is *never* wrong, else it would not be *intuition*. Now, for many it is difficult to become accustomed to follow their intuition, because it only makes itself known during a micro-second of your time. If you are busy in your hurry-scurrying day-to-day life and you are making decisions after decisions and in that micro-second a voice will say: 'no' or may say: 'yes, do this' or 'that' and you are so busy and you are so out of tune, then you do not listen. *What you call intuition is feminine energy.* It would behove you all very well to listen to the little voice inside, hm? To come in touch with this is very much - as we have spoken before - like listening to your body; when your body is speaking to you, when there is a dis-easement of your body, then it is telling you something, eh?

Now, in this matter of feeling, humanity is terrified to feel. A common bonding of humanity is called pain, is called anguish, is called desolation, is called broken heart. From the time of your birthing, from the time you first experienced the anguish of invalidation, so each and every one of you has built a wall of invulnerability around yourself. All, all of you are afraid to be hurt yet again. The disappointment, the grief of living - pain - all of you know it very well. You all discover it when you are very young and you confuse pain with feeling, but you see, beloved ones, *pain is not feeling*. That which is *pain is resistance to feeling*. It is important that you understand this. *Pain is resistance to feeling.*

We will explain it thusly: When situations occur in your life that are of great joy, the feeling centre, which is in the belly, is open. When you have great laughter it is from the belly, hm? - and when you judge the situation to be of joy or of great laughter, the energy centres within the body are open. Feeling is energy, that is all it is. That is all *everything* is - energy. So when a situation arises that you judge to be joyous, all of the chakra energy points within the body are open. And so the joy starts in the belly and travels without resistance through all your open energy centres and your heart becomes alive. In that moment of joy you are living in the now moment of your time, without past and without future and every atom and molecule in your body resonates with that joy. *Beloved ones, joyous people have no diseasements of the body.*

Now, when a situation occurs that you judge to be painful, that you are in fear and trepidation of - that it will bring pain - then the energy centre in your belly is closed. Each and everyone knows that when such a situation occurs and you are shocked and in pain, where do you feel it, hm? In your belly! We have heard the expression: 'Kicked in the belly by a mule', indeed? Judgment becomes like steel claws within the belly and the energy has nowhere to go - *that* is called pain. Now, what do you think creates pain? *Judgment. There is no other cause of pain.* It is the judgment that creates the resistance called pain. Pain, beloved ones, is accumulative. You have known it for eons and eons of your time, lifetime after lifetime. You are all very familiar with it.

So, what occurs in your life? As a small child you are invalidated. It need not be anything of a 'big deal'. Something quite simple. You are smacked on the behind area and told you are naughty, when in your heart you were not. You are told that you are not beautiful, you are told that you are stupid and much, much more and so you judge yourself to be not enough and as you judge yourself to be not enough you experience pain. As you grow you learn to 'deal' with it and what does it mean, to *deal* with it? It means that you build a wall of invulnerability around you, so that you will not be hurt again. But you are and so you build a higher wall, a thicker wall. And as you grow to adulthood, the wall indeed becomes very large and very thick, and you learn not to show your feelings. You learn to hide and fear to show your heart, because if you do, you will be wounded. Each time you are pained, you 'get over it' and it occurs again and it becomes more and more and you are more and more in judgment of Self, in judgment of the rest of humanity. Very often you are a 'victim' of governments, 'victim' of the persons in your life and as you are more in fear, so you are drawing more and more and more to situations and people who give you exactly what you believe. So, do you understand how it becomes a circle? What you are all terrified of is to be vulnerable and you are terrified to say: 'I am afraid' or 'I am hurting'. You are afraid to say: 'I live my day-to-day life, but in truth I am dying of a broken heart. And you see, beloved ones, all of you are dying of a broken heart that you cannot speak of, that you are afraid to look at, afraid to acknowledge who you are.

You are all so afraid of not being enough and you do not dare to say so, because you think that everybody else has the answer, that you indeed are worse than everybody else in your secret heart. You look at all the areas in your life, your secret thoughts and your so-called hidden behaviour and indeed, you 'know' that you are not worthy, you 'know' that you will never 'get there'. For many of you it seems like a charade *and all you truly desire is wholeness.*

And all of this is because you do not truly understand that who and what you are is divine expression. Who and what you truly are is so wondrous beyond words. When we say to you, indeed you are the thought of the All That Is, of the Source, then what you do is to personify that which is the All That Is. When we say God/Goddess, you do indeed imagine that there is a personified God, but you see, it is showing lack of understanding about the nature of First Cause. It is truly not a personification and *when you can imagine that the All That Is has no personification, then you may truly see the All That Is in everything, including who you are.* The divine thread is in every atom and molecule, in every universe. *The force is within everything* and you may look at all things and *know* within your heart that every blade of grass, every bird and flower and tree and creature - every rock and even that which you consider to be man-made - all of it, all, all of it is indeed the All That Is. When you can understand this reflection outside of yourself, then it may help you to understand, that, indeed, this is also who you are. Sometimes it is much easier for you to get the picture outside of yourself, than it is to get the picture about who and what you really are. And after all, all we really desire is that you do know who and what you really are.

Very well, dear ones, at this time we would ask you for questions, if you have your great notations there within you, hm?

Q: (F) Thank you, P'taah. So what you just have been saying could also be interpreted as there being no model for perfection. We have a long history of thinking that there is a model of perfection.

P'taah: It is a very good question. Now, we have spoken to you before in your time about role models in your cultural society. Perhaps it bears to repeat. You see it is part of you being unworthy, because there has been cast for you an idea construct, which bears no relation to reality in truth. We are speaking in terms of universal law. There has been - not only in this culture, but in many cultures - that one persona with the embodiment of all wondrousness - perfection - and so all of you truly, emotionally aspired to this role model and we will use in this instant that which is Jesus, the Christed One. We have indeed noted that this is very prevalent in your culture and as we have said before, whether or not you intellectually agree with it, we are speaking of your emotional aspiration to this role model. We are speaking of the collective consciousness of the morphogenic resonance of your culture, which you are tapped into, whether or not you understand that.

Now, that which is the Perfect One, that you are all supposed to be like, to aspire to, you see, creates an incredible dichotomy, because what you understand to be perfection is called a finished product. Indeed? You are not [finished] and never will be and neither is anything else in this dimension of reality and, indeed, in any reality, because once a thing is finished it is no more. Do you understand? Everything is in an impetus of creation, is in a state of growth, of expansion. Everything, everything. Every cell in your body is bursting with creativity, it does not stay still. There is not a molecule which stays still, else it would not be. Looking at microcosms and macrocosms - does anything in your universe stay still? Of course it does not, and so that which you believe to be this finished perfection, that you are all striving for, *is nonsense.* You think that when you are enlightened masters that it will all finish? I will tell you: it will not! - because the dimensions of reality are infinite. The energy of First Cause does not stay still. It is constantly expanding, moving, else you would not be here and neither would anything else. Do you understand? So indeed, it is for you to look at, to have an understanding, that what has been set forth for you as a role model indeed has been a wonderfully effective tool of enslavement. *Let there be no mistake: That which you call religion on your plane has always been, and is now, a tool of enslavement to keep you in chains, to keep you in control that you may not know sovereignty and free dominion.*

Does that answer your question, beloved?

Q: Yes.

P'taah: Indeed.

Q: (M) It must be very disappointing for Yeshwa Ben Joseph to see what we made of his teachings.

P'taah: Not really, beloved. That which you call Yeshwa Ben Joseph indeed now is an *idea construct.* What people believe about this one, who was indeed in physical reality, has nothing to do with what was the persona. It has become an idea construct and the idea construct indeed has a validity and reality. It is reality, but is up to you to choose *not* to partake of this reality. There is no judgment, beloved ones, outside of here. Whether it is your 'here' or somebody else's 'here', hm?

There is no judgment in the universe, there is no Yeshwa Ben Joseph up in the clouds, there is no God in the clouds keeping notes on you and saying: 'You will not make the grade'. It is your reality, it is for you, day-by-day, to understand that you draw to you exactly what you need to learn. And as you learn you need never experience again [these lessons]. Questions?

Q: (M) *So judgment seems to be in the way of living in a greater harmony. So, how does one leave judgment behind, for it happens all the time?*

P'taah: Let us not also confuse, beloved one, judgment with discernment. Do you understand the difference? Judgment is when you say: 'This is right and this is wrong'. 'This is good and this bad'. Because in truth there is no right and wrong or good and bad, there simply is - there simply IS. Now, something else we have also noticed - particularly among the 'New Age' thinkers, you know who they are, hm? The 'New Agers' - what is the word, beloved one, *(P'taah addresses a lady-friend of Jani King)* the word my woman uses for New Agers - 'whoofties'? Now, we are jesting, because indeed, the New Age is so, it is valid - it is a designation of those who are eager to participate in an expansion [of consciousness]. However, we have noticed that many who are involved in the New Age movement have taken all the sanctimony from the hypocrisy of their old religions and simply transposed it to New Age thinking. That which is the sanctimonious - the holier than thou attitude - is hardly expansion and we have noticed many of these dear people who say: 'This is wrong and this is right and if you do not do it this way, then you are not of the New Age and you will never get there.' 'Oh well, this one is too far from enlightenment' - 'This one is a great warrior, he will never make it', hm? Well, that is alright. Good, bad, right, wrong - basic judgments. It is not easy to overcome that which has been in your culture for eons of time. Lifetime after lifetime you have known good, bad, right, wrong - that is alright. Discernment, is when you can say: 'This is one form and this is another and this is my choice'.

You know, my woman often says: 'Hm, what a bastard!' and people are very shocked. How can someone so involved with this wondrousness, say such a thing. That is not only what she says, either. She can become very colourful. *(Laughter in the audience.)* However, never mind about the word, it is called intent and there are many who indeed mouth the New Age platitudes, whilst in their hearts there is raging judgment - and that is alright. But it is merely for you to be aware and as you are aware so you will come into an understanding of what judgment is.

Do you know, dear ones, the greatest thing that occurs with people who are chasing enlightenment? They know that judgment is a no-no, judgment is 'wrong' and so they are in judgment of the judgment. 'You cannot judge, judgment is wrong, that is not enlightenment.' Is it not silly? To make judgment wrong, beloved, is to get straight back on the old track again. So it is not to be silly about it. Judgment is valid, because it exists. Judgment is a divine aspect of who you are and the only way that you create the change is by embracement - by acknowledgement and the acceptance that who you are is judgmental. As you desire the change and as you embrace that part of you that is judgmental, so will you create the change, but what you invalidate, you will empower. And truly it is to remember, *what you invalidate, indeed, you empower,* and that is exactly what you will draw to you, if you invalidate it.

You see, it is all a great paradox, a dichotomy. Well, there is no point in being too serious, because, dear ones, I will tell you this: The greatest aligner of all is laughter, when you are truly laughing from the belly. Indeed I would say that there are many things in this life and certainly in all of this very serious business that is very funny. And as you laugh and find joy in it, even in the paradoxes, so will you be living in the moment, without past and without future and create joy in this fashion.

Now, at this moment, dear ones, we will take a break, so that you may refresh your bodies and then we will return that you may ask many more questions, indeed.

(P'taah takes leave and returns after the audience had a break.)

P'taah: And so - questions, dear ones?

Q: (M) It is about laughter and that laughter aligns. On your level of consciousness - can you too be heartily childlike and silly?

P'taah: But indeed. Childlike, you see, it is not childish, in the connotation of what that is. Childlike in the sense of wonder, do you understand?

Q: Yes.

P'taah: And so in that sense of wonder all things are made new, In that sense of wonder is the recognition, and indeed the awe, with which you behold your universe; in your recognition of the divinity of all things.

Q: (M) What about funny cynicism?

P'taah: Hm, indeed dear one, it is alright, you know. To be cynical, to be knowing about the foibles of mankind and to understand, dear one, that as you are laughing at the foibles of another it is truly that you are laughing at yourself, hm? There is nothing wrong with that.

Q: But you laugh at your level!?

P'taah: But of course, do I not laugh at this level? We would hate you to think that we are very serious, that we do not enjoy laughter and jesting. It is all part of joy. We have noticed in another sense also that for many, laughter is a measure of desperation, and certainly if you would not laugh then you would cry. And we have noticed that laughter is indeed a great deflector of tears. But you know, whatever the emotion is, when felt grandly, it means that you are [living] in the moment. All is valid, but it certainly would be our hearts desire that the great emotion is that of joy. - Questions?

Q: (F) You were speaking about the memory of painful emotions which we pile up inside, and I would like to know where the value of emotional release work comes into this and what is the right attitude toward it?

P'taah: Now, in this fashion you are addressing that of which we will speak of again in more detail. But we will say this for the moment: There is, as I have said before, the terror humanity has about feeling. Now, when the energy of feeling builds up within the body, there is expression and that which is suppression. When feeling is suppressed, that is indeed when the body comes into great diseasement - the suppression of feeling, the non-allowance, if you like, of expression, hm? There are many new therapies and when we are saying new, we mean this recent age, where people are encouraged to express, where people are going back in their life to re-experience what has been of great pain and to give forth expression to that. And that is very good, it can be likened unto 'taking the pot off the boil'. However, expression and suppression are the polarities and within the middle of the two lies your answer and it is called acknowledgment and embracement. With the acknowledgment and the embracement comes what is laterally termed the transmutation, which creates the change from pain to joy. However, beloved, if you bear with us, we will speak in very great detail of this very soon. But for the reason of producing this manuscript, we are taking things slowly and a little in order. But you will be with us, beloved, you will not miss out, indeed.

Q: (F) I just want to ask you something about dreams. You said along with laughter goes imagination and feeling to flow energies free through the chakras.

P'taah: Indeed.

Q: Imagination. When dreaming, I sometimes experience my daily details over again and a lot of the things I have done or thought of doing or I am yet to do, I am experiencing all this in my dreams. What is this when we sleep, is this our imagination or what is it?

P'taah: It is all of that and much more. Your dreamstate is multidimensional reality. It is much more than merely putting in order the events of the day. It is much more than experiencing your imagination. It is on many, many levels, because you also travel when you are dreaming. When you are in your sleep state, you are at one and the same time on many levels of reality. That is, the whole SELF - the soul energy, if you like - comes together very often with other parts of the soul energy, which are inhabiting this planet in the same time and space frame, to have communion. There is also the travelling forth in super-consciousness to that which be other realms. These may be other planets, they may be other states of reality, which are beyond time and space in your comprehension. And do you know, it is real, it is valid. When you are in this dream state you are learning other lessons and indeed, very often those experiences are beneficial to your life in this reality and are truly lessons, learnings or understandings beyond the comprehension of your present consciousness. Therefore, the conscious mind creates pictures for you that you may be in conscious understanding. Now, we have said before that at this level the consciousness, if you like, grabs on to these pictures. It is beneficial, and we have stated this in the days before, that you would create for yourself a book that you have beside your sleeping place and when you awake from your dream and it is still very clear in your mind, that you make a note of your dream. It does not have to be very elaborate, perhaps a key word or a few sentences so that you may remember - and you will. And in this fashion, if you are doing this night after night, it will create a pattern for you to see with your consciousness, so that you may understand on a conscious level what you already know on an unconscious level, do you understand? And so in this fashion it is beneficial. However, there are truly things that you do experience in your dreamstate that are beyond consciousness and this time - in other realms of reality. Dear ones, in those other realms of reality it is indeed *this* state which is the dreamstate. It is a wonderful playground for you. For all, this dreamstate, dreams, imaginings, playing out realities that seem so distant for you, are nevertheless as real and as valid as this day-to-day reality, that you are in conscious cognizance of. Alright, dear one?

Q: (F) I understand that we are of the divine, but is there a key that can help us understand who and what we are?

P'taah: Every moment is your key for that, truly. You know dear one, it is so simple, really, and that is the dichotomy, because it is so simple and yet it is truly so difficult for humanity. What is the key? The key in truth is that you learn who you are. Learn what you believe in. Your belief structure is your key. When you understand what you believe, and you realize that much of it is nonsense, and that you did not even know that you believed such and such - and when you can acknowledge and accept the layers upon layers of who you are and you can indeed embrace every facet of who you are - so you will come into recognition of your own divinity. *So you will know the GOD I AM.* It is truly that simple. Having said how simple it is truly, we understand that it is not so for you and so we take a step at a time in bringing forth that which you already know. Thus it becomes amplified within you, that you may come into the knowing. At least, beloved, that is what we are attempting.

Dear ones, enlightenment is not about being good, enlightenment is about *Being. (And P'taah repeats, softly:)* Enlightenment is not about being good! It is about Being.

Q: (F) My question concerns sleeping and dreaming. Often I awaken with a suddenness, without apparent reason, can you explain this for me?

P'taah: Everybody does. There are several levels of this. One is, that very often, when the body is in relaxation, some muscles experience spasms. This is one of the more mundane reasons. The other is very often that the vibrations of the body are speeding up, often as a prelude to consciousness leaving the embodiment. Sometimes it is that the vibrations themselves will wake the body because the consciousness will know it is not time to leave for one reason or another, and it is the coming back into the body which creates the waking, because it is like a shock. But always with these things the answer is more than one thing, because of the different levels of occurrences within the body itself and also because of the relationship between consciousness and the body. They are very closely linked, as you know. That which is the journeying out of the body is what you all do, always - though most often you do not have a conscious recollection of your journeys and adventures outside of your body. Some of you do have conscious knowledge and indeed travel forth from the body, when you are in the waking state and not in the sleeping state, but it is almost a state what you would call hypnagogic, which is a state just before you are sleeping and when the body is in total relaxation. Sometimes, when you are practicing your meditations, your body comes to this state, which is occasionally referred to as alpha state and in this state people will very often experience journeying out of the body.

Q: (F) Referring to something you mentioned before the break: When we do remember to take our pain more lightly, it seems to be easier for those of us who have fortune in their lives, yet when we are faced with the death, disease and disaster of others, apparently

less fortunate than ourselves, it is tempting to feel guilty about having fun while these others are suffering.

P'taah: Guilt - responsibility. Guilt is a lesson not learned. What is responsibility, beloved?

(The lady questioned by P'taah hesitates with her answer for a moment; the situation is tempting a gent across the room to 'rush to her aid' and he whispers: The ability to respond to ...', but P'taah swings around and cuts him short, feigning the manner of a stern school teacher:)

P'taah: We are not speaking to you, beloved.

(The audience bursts into laughter.)

P'taah: We know that you have top marks in class.

(The entire audience is greatly amused by the little intermezzo and hearty laughter bubbles up again. The lady, helping the man over his embarrassment, very gently says:)

Q: He only tried to help me remember..

(The gent, feeling much like a little schoolboy caught in the act, is apologetic, looks at P'taah and says:)

Q: I am truly sorry.

P'taah: Do you feel guilty?

(The gentleman, mischievously smiling:)

Q: No!

(The audience roars with laughter now and after everyone has settled again, P'taah continues:)

P'taah: Responsibility, my beloved children, responsibility is to respond to that which makes your heart sing. Responsibility is not duty. When you truly do what makes your heart sing, that is responding to who you are. Then you make the world happy, You are not responsible for anybody else's happiness. You are not responsible for anybody else's life. You are only responsible for your own life, because *life is a co-creation of all beings.* Now, there are many who say: 'But, P'taah, what can we do, when our desire does not coincide with that of our loved ones?'

Well, beloved ones, when you are loving somebody, then you wish joy for them - when you love your Self you are wishing joy for yourself. When decisions are made in love of Self it creates harmony, well being and joy. If you are doing something not out of love but out of duty, then what is occurring is resentment, and do you think that people do not know? It does not matter how much you smile and say pretty words. If there is resentment of duty in your heart it is felt by all. And if you speak forth from your heart and say: 'I understand what you desire for me, but truly from my heart I desire to do such and such - and it is not because I do not love you, because indeed, I do - but I must live my life in joy.' If somebody takes it ill, it is not your responsibility. It is their creation and *what somebody else thinks of you is none of your business,* because everybody responds according to their own fear and their own judgment. You see, if somebody died, if it is someone you love very much, then you are indeed in grief, because you know the physical contact in this fashion is at an end. You, dear one, are also knowing that if someone you love dies, there is no end for anything, because in truth there is no death.

It is very fashionable that you have great mourning. It is less so in this time and space. But it has been a great tradition, that you do not do so and so, because of mourning. If the one you love loved you indeed, would they wish that you go around with a long face - worse still - that you are pretending? Truly you are to dance and sing, hm? In many of your old cultures, when one was to make a transition from one reality to another in death, it was cause for great celebrations. And in truth, the grief and the longing to be conjoined once more is perfectly natural, but it is always to remind oneself that there is never an end to love. When you have loved somebody, that love goes on to infinity - there is truly no separation.

Q: I was particularly thinking of people in other countries, who are less fortunate. If I am thinking of their death and disease as wrong, then I am actually in judgment of it.

P'taah: Everybody chooses their own reality - at soul level - for what ever it is that they are desirous to come into understanding of. We understand that when you are caught in this dimension of reality, you say: 'Why would anybody choose to be born, only to starve to death as a babe, to be born without all physical faculties, to be born into a time zone of war, destruction and desolation - to be born into a family of violence, of hatred - why would they choose such a thing?' But you know, it is all an illusion, it is a movie that you are creating for the experience. You know, it is not that life is a 'one shot' affair. You may rest assured that if you are born into a situation of harmony, of abundance, of great joy - mind you, I must tell you that there are not many of you who have all of this unendingly - then you have been the exact opposite in other lifetimes.

We have said before: you have been the oppressed and the oppressor, you have been the murderer and the victim. You have been of grand social standing and of very lowly states, you have been man and woman - you have been everything. You are every facet of every being ever to walk this planet and more. And so you create your reality for the experiences. And it is all a grand illusion.

Q: (M) P'taah, could you just explain to me, how we best enter the state of being?

P'taah: Dear one, I am very happy to speak of it. To be a human being - not a human doing, eh? *(The audience chuckles and P'taah states that the remark is not original.)* To live in every now moment in the fullness, that is what we are saying. Not to be so engrossed in your past and to be so worried about your future, that you do not live life now. How do I tell you how you may be more still? How you may be less in judgment? How you may be less in expectation, but rather in expectancy? How you may be like the dolphin - like the child?

Well, we would say it takes practice. But you may catch yourself being engrossed in your past and engrossed in the worry about your future. Then it is simply to say: 'There I am again' and to say: 'But what am I doing *now?*' Little by little it becomes easier for you and it is not to worry and it is not to judge yourself. It is merely to catch yourself sometime and as you do so, you may say: 'I am so busy thinking about the future, now I will simply *be.*' Is this of assistance?

Q: Yes.

P'taah: It is not in your education to be in the now moment, because in this society you are taught, from the time of your childhood, to think about your future. As you are growing, very often among humans there comes a bitterness about the past, because you do not understand the lessons to be learned. You hold on as you do with guilt, but when you learn the lessons, when it is integrated and embraced, why indeed, then it may be relegated to that which is not of importance. In this way you may change your past, that is how powerful you are. We will speak of this more, when we speak to you about transmutation, about how you change your past and your future, because, you know, it is all the same.

Q: (M) *This evening, earlier on, you said thought is much more than a tool. I sense that there is more to it, but could you elaborate on this?*

P'taah: Thought is truly what creates your universe. So, indeed, thought is a tool of manifestation. What occurs in your reality may not occur without thought, do you understand?

Q: *I think I do.*

P'taah: Well then, if there is a war, it cannot occur without somebody thinking of it first. Before anything can manifest in physical reality, in third dimensional reality, it must commence with the thought. You are the result of thought. Physical matter is consciousness and consciousness is in truth thought, but thought is also a source of power, and we are speaking of a technology which is beyond of what is known at this time. It is not appropriate at this time to go into this in much detail. We have touched this briefly before, that in a grander technology thought is used as a power, as fuel if you like. For example, your grand computer is powered by electricity. In a technology to come the computer is powered by thought.

Q: *That would explain my feeling that there is more to thought.*

P'taah: Indeed, dear one. Your scientists, that is your physicists, are now coming into the understanding that it is thought which creates matter, and indeed, that who you be is called a thought of the All There Is. Thought is of a higher spectrum of vibration than light.

Do you understand? Once upon a time your physicists thought that there was nothing of a higher vibration than light and, so indeed, that nothing could travel faster than light. *This is not so.*

(A young lady becomes intrigued:)

Q: If thought is of a higher vibration than light, then we should be able to move things through a single, concentrated thought, even teleport ourselves.

P'taah: But you can, indeed.

Q: So why is it that we cannot, when we try to do so?

P'taah: Because you do not understand about pure thought and *because you do not believe that you can.* You see, dear one, your belief structures are the foundation stones of your reality. We have said many times before and, indeed, we will say it many more times, that if you desire to walk through a solid wall, then you may, but if you believe you cannot, then you cannot.

(A lady seeks clarification on a different matter:)

Q: P'taah, could you explain to me why all the spiritual masters of the past, Jesus and Buddha for example, and beings like yourself are of male gender?

P'taah: Now, why is it that all of this is masculine and not feminine energy? It is not so. As your history is rewriting itself towards what is called a masculine bias of energy, female of gender has been overlooked, has been indeed hidden. That which is masculine has been very afraid of woman, because for man in your history it is woman who has been all-powerful, because woman is very close to that which is Goddess, that which is your Earth. Woman has been the creator of life. In this epoch of your history, the masculine, the male gender, has always been terrified of her power. That is why subjugation of women has been so horrendous. Now, it is certainly changing. That which be the entities, who speak forth in this fashion, such as we, are really not of masculine energy in truth, but indeed that which speaks forth in an appearance of masculine energy becomes easier for people to relate to, than if *feminine* energy were given forth. In the consciousness of humanity it has been masculine energy which has been of authority, which has been of knowledge. Do you understand? And so, in this fashion it is that masculine energy is being called forth.

Q: So, because of our belief systems, including mine - because I respond better to masculine energy - we are calling forth masculine energy?

P'taah: Dear one, it is that woman has been in a subjugated position, because man is afraid of being powerless and so has put forth power and certainly does not take kindly in your culture to admitting that all knowledge and wisdom would come from woman. But you see, in truth, that which be this energy is not so masculine.

Q: Are you implying androgynous energy?

P'taah: Dear one, that which is the soul has no gender.

Q: Is it beyond gender?

P'taah: Indeed, incorporating all. Your soul has no gender, Indeed, you have been man and woman lifetime after lifetime. It is not really that it is beyond - it is just that it is incorporating all.

Q: If we can choose to come as man or woman in each life, why is it that we have homosexuals and lesbians? Why do they not come back as women or men?

P'taah: There are many reasons. As we have said before, humanity is really bisexual, that is to physically extend love, no matter what the gender. Now, we understand that this is certainly not so popular, that there is great judgment about those who are homosexual. However, we would say this also: There are occasions when it is almost a little bit of a mix-up, and we are speaking of the physical embodiment here, what is called the chromosome. Most often homosexuality is created by trauma. We are not speaking of the sexual expression, when we speak of the soul energy having no gender. There is no judgment of homosexuality. We have heard in jesting, whether one is homosexual or heterosexual, they are still confining themselves to one or the other, when in fact they may have a physical expression of love and joy amongst all. *Sexuality, you know, is not merely meant for procreation.* That has been another method of enslavement for humanity. If it were merely for procreation, then it would, perhaps, not be so pleasurable, eh? We understand this is a very touchy subject for humanity.

Dear ones, we feel it is sufficient unto the time - that is, the absorption level is at saturation point. It is indeed with great love that I give forth thanks and honour to you all. As always it is joy to my heart to be with you and sadness indeed, to be saying farewell for the moment. However, we shall be together very soon and it is also for you to know that when I am not with you in this fashion, you may indeed, in the quietness of your own time, speak forth to me and I will be there for you. Indeed, beloved. Our thanks to you all. And so we will bid you a good evening and bid you go forth in love and in joy - in light and in laughter and to know, indeed, that I love you. *(The audience assures P'taah of their love for him.)*

P'taah: But I know!

Chapter 4

FOURTH TRANSMISSION.
Date: 18th of September, 1991.

(P'taah greets his audience in his usual dynamic manner:)
P'taah: Beloved ones, greetings. A very good evening to you all.
(Looking at new-comers to this event, he recognizes among them persons he had seen fourteen months prior to this evening.) Well, indeed, so - old friends again, and new ones. You are all well come. That which you have called forth is to raise the vibrational frequencies of Earth's humanity. Now, we are not suggesting for one moment that this is your conscious desire. Indeed, we understand that the conscious desire of each one of you, is that you come into the knowing. However, at the level below your consciousness or, indeed, above your consciousness and certainly behind, there is the fervent desire of all of humanity to come into harmonious vibration with the totality of the All That Is, the First Cause - God/Goddess if you wish. However, as we have said many times before, very often we are not speaking [in terms] of God/Goddess, because for humanity there is always a personification of the God/Goddess, of the All That Is. *As you put forth a personification, you are limiting the Source.* When you personify the Source then, indeed, you are very inclined to forget that everything upon your planet and in the multiverses is a reflection, is an expression of that Source. Now, what you are is vibrational frequency. Thought is vibrational frequency and what you all are is the thought, if you like, of the Source. What you are is a vibrational frequency very often not in harmony with the vibrational frequency of your universe. So it becomes discordant. It is not in harmony, because, beloved ones, you have forgotten who you are - you do not resonate to the harmony of divinity and yet, what you are in truth *is* divinity, of course.

We have said it many times before, and in truth we cannot say it often enough so that it becomes part of your belief structure: *you create your own reality absolutely*. It is your belief structure, your consciousness, which creates physical reality and indeed every other one; but for this moment we are speaking only of the physical reality. What you believe about yourself is reflected all about you. Everything you see outside of Self is indeed a divine reflection, a mirror.

Now, when you are in harmony with SELF, all things are revealed to you: all knowledge, that what you all dream of - intergalactic travel - the knowledge of the multiverses. We understand humanity's fervent desire to fit together the pieces of the puzzle. Each time we say to you, dear ones, the answers to the pieces of the puzzle, all the solutions, lie within you, not without. It matters not how many books you read, how many degrees you have in your physics, you will not have the answer, because the answer lies only in the resonance of divinity. There have been many civilizations upon your planet, both in your period of history and your pre-historic past, grand civilizations, who have had this knowledge and who have chosen - for one reason or another - to go forth from this place.

Now, we have been speaking in these weeks very much about your belief structures and what you believe about yourself and how this is reflected in your life; how you build upon these belief structures and how, from your early childhood - indeed from the time you were a babe - you view your world through a lens. How, from the moment of your first invalidation and from the time of your birthing - that is from the moment you tap into the morphogenic resonance, the collective consciousness of humanity - you build your life accordingly. We have spoken very much of how humanity lives with a broken heart, each and every one of you. Dear ones, it is true: You are dying of broken hearts. Each and every one of you lives with desolation and anguish from the time of your childhood.

Now, you say to me: "It is all very well, but what do I do about it?' Well, tonight we speak of it again - and some of you have heard it before and for some it is new - for all of you it is the most important part of your development: *Until you understand how to embrace that which you judge to be painful, you will never have your energy centres free to tap into all the knowledge that is within you.* And so, dear ones, this evening, we are going to speak about transmutation and from thence we will speak of it again and again. Transmutation - it is called alchemy. You know what alchemy is? Creating change in molecular structure. So it is, that emotional transmutation is what changes agony to ecstasy. And what indeed is ecstasy? Well, *it is divine Oneness, it is lack of separation, it is when the Self is totally integrated with All else.* Divine ecstasy - wondrous indeed. It is a miracle, because not only is it the alchemist's art of transmuting agony to ecstasy, it also creates molecular change within the cellular structure of your body. It creates a change within the brain. It changes what you term to be your past. It changes your future, and indeed, it changes the face of your planet. A miracle indeed.

Now, dear ones, we will say this to you: What I am speaking to you of is simplicity. It is not something that you may wish to attain but think 'I will never make it,' and you are all very good at that. It is not a pie in the sky. It is not like hearing about ascended masters and saying: 'Hm, I wish it were me'. What I am speaking of is for every single entity on this plane. And truly, *it is not beyond any of you.* And we would not wish that you would worry that you do not 'get it', because *you will get it* and you will be able to practice. Now, we will recapitulate:

From the time of your birthing you have been invalidated, you have been hurt and you have been wounded and you have been judged. It is from this that you build the picture of yourself and of reality. So by the time you are young adults, you have already formed the belief structure about who and what you are. All, all of you believe that you are not worthy, that you are not enough. It does not matter what face you put on for the rest of humanity. We are not speaking of this. We are not speaking of your grand courage and brave front, we are speaking of who you truly are. That sense of unworthiness is certainly a common factor of humanity. Now, the way you judge yourself is often so painful that you cannot even bear to look at yourself, you do not know who you are, therefore you place all the judgment outside of who you are. We will not go into this now, because we have already spoken at length about judgment and we will certainly do so again. However, for the purpose of this manuscript we will say merely this: As you judge yourself to be not enough, as you judge yourself to be not worthy of love and as you cannot love who you are unconditionally, as you cannot accept and acknowledge every facet of who you are, so it is reflected in all facets of your life. And so as you are hurt and in pain, not only with the judgment and invalidation of other people, who merely reflect your own judgment about yourself, the energy centres are not free and you live in pain. Pain indeed, as we have said before, is the resistance to feeling. *Pain is not a feeling - it is resistance to feeling*, it is judgment - *pain is caused by judgment*, it is only resistance. We will say it again and again in these weeks. *When you are in judgment of situations in your life, allow the feeling of joy to permeate your being*, then there is no resistance, the energy channels become open and the feeling is allowed to travel from your solar plexus and up to your heart and indeed to the crown and, in this fashion, resonates through the universe and touches *all* things, *all* consciousness. When you judge that you may be hurt, when you judge the situation to be 'harmful' to you, immediately resistance comes like claws into your solar plexus, thus closing down the energy centres. The energy has nowhere to go and so you are in pain. Very often this is physical pain *as well* as emotional pain. Now, when a situation occurs that you judge to be painful, your normal reaction is to build a wall to hide behind. As we have said, the wall in your life becomes higher and higher and thicker and thicker, so that you will not be hurt, but it makes no difference at all, of course you are hurt.

So, transmutation - we shall give you very simple steps. The first step is to take *responsibility*, because you create your own reality absolutely. There is no situation, there is no thing in your life, which is not co-created. So, responsibility: *'I did it'* - that is not so easy. It is wondrously easy, when a situation brings you joy. It is not so easy, when it brings you an ache of the heart.

Step number two: *Judgment.* When you judge somebody who has done a wrong thing to you, what do you say? 'The bastard!' Hm? We hear you! We hear our woman - also. So, *the moment you blame, you do not accept responsibility.* When you have co-created something in your life which is not so wondrous, it is for the learning experience. Everything you choose, you choose for the experience, that is all, for the experience. Within each situation there is a pearl, a jewel, as you, dear ones, are jewels. Still not so easy, hm? How do you bless somebody who has caused you great pain? Nevertheless, it is indeed to say: 'This I have created with this person or persons.' And then you say: 'Oh, my God, I am in grave judgment.' But indeed, dear ones, that is alright. Judgment is a valid and a divine expression of who you are. It is to embrace the judgment and to bless the persons, yourself and the pain. And know that the pain, which has accumulated over eons and eons of your time - lifetime after lifetime - you have accumulated it so that you may discover the pearl within. So it is to embrace all, to acknowledge, to accept. Embrace all into the light of who you are and as you do such, the claws of the pain, which are suppressing the energy, are relinquished. Thus may occur transmutation. To go from the belly to the heart and to the crown, then you will know ecstasy. Then you will truly understand *Oneness.*

You know, there is nothing to do, beloved ones, you are not to *do* transmutation. It is not possible, because as you are doing it, is called masculine energy. It is called striving. Transmutation is only to be *allowed.* Allowance, acceptance. It is created through the nurturing of who you are. It is called *surrender.* Now, surrender, hm? For very many of you it is really a dirty word. Because in your culture surrender is not noble, it is not heroic. I will tell you something, dear ones: *Surrender is to join forces.* Surrender is to acknowledge and accept who you are, and dear ones, who you are in truth is divine expression. Until you can come into the idea that who you are is divine, else you would not be here - until you truly can understand that every facet of you, every thought you have ever had, every action that you judge so harshly, is truly divine expression, else you would not be here - until you come to know that all things in the known and unknown realities are indeed a thought of the Source, you will not know that you are absolutely God smelling the rose. That is the ecstasy of third density.

So, dear ones, you have many questions - let us begin. — No questions? Time for me to go.

Q: (F) Is it possible that you are specifically connected with Jani's consciousness or are even a greater part of her consciousness, and that we all can contact and communicate with a greater part of our own consciousness to learn in a more conscious way about our being? Also are extra-terrestrials and other beings, that are appearing more in this dimension different aspects of human beings?

P'taah: Was that all one sentence?

(Amusement all around.)

Q: No, two sentences.

P'taah: Very grand indeed, beloved. Now, there is a very simple answer: *There is no separation.* There is indeed no separation between that which you regard as your consciousness and that which you regard as your super-consciousness. There is no separation. There are certainly different vibratory frequencies. After all, everything in your universes - multiverses indeed - is a vibrational frequency. Now, in regard to this energy that is speaking with you and that which is the consciousness of our woman: We are of the same vibrational frequency, but that who I am, is certainly not only contactable through our woman, as you know beloved, because you have spoken with me. So we would say to you that it is very important that little by little you come into the conscious knowing that there is no separation. Even one to one, whether you judge it to be lovers or enemies - truly, if you understood that there is no separation, there would *only* be lovers. The whole world would be lovers, indeed. Then we would come here forever in our body. We do love lovers, you see? Questions?

Q: (F) With some people we seem to have a life-long relationship circling around certain lessons or problems, where we repeat the same unsatisfying issues again and again, which obviously leaves a big question mark for me. I wonder what I can do to complete the learning?

P'taah: You know beloved, this is a very good question, because for most people in adulthood, when they glance back over their life, indeed it is a succession of repeated situations. It may be a different story, but the feeling is the same, whether it is with one person or whether it be with many. Relationships are indeed the greatest lesson of all, because it is in that fashion that you understand how it is within. It is to look at the feeling engendered, the story does not matter. Beloved ones, it would behove you to remember that indeed you choose those who you love most, lifetime after lifetime, to learn your lessons. Now, *reality is thought embraced by feeling.* We have spoken to you about how to recapture something wondrous in your life and we are speaking especially of those moments of Oneness. It is not to remember the physical aspects of 'it was a beautiful day and the sun was shining; I was by the seaside and there were many flowers', hm? That is called outside dressing. The scenery, in this context, means nothing in truth, for if you tap into the feeling engendered, then you have recaptured the moment. When you are considering your relationship - and it is not the only one - then you will find that there have been many times when what is occurring creates the same feeling within you, do you understand?

Now, I have just now given you what you may do with it *[the recurring unsatisfying issues]* and then you will change the external reality. We will say this to you also: It is important for you to remember that all of you have a very great investment in your pain. You have a very great investment in anger, in aggression, because you know it so well. It is beneficial to remind yourself that everything, *everything which is not an expression of love is an expression of fear.* When you find yourself in situations wherein the fear is inflamed, which will create anger, hostility etcetera, etcetera - it is to know indeed that it is only fear that is to be embraced. *As soon as you invalidate what you feel, you empower it.* If you are invalidating hostility and anger, if you invalidate impatience or cruelty - whether it be your own or whether it be the mirror of somebody else you are looking at - you empower that. So it is not to suppress, it is not to drop, to push away, it is to embrace, dear ones. It is to take every aspect and know it is all part of who you are. It is valid, it is divine expression, only awaiting embracement to create the change - to embrace it into the light of who you are and as you do so you are creating more and more and more *light.* Harmony - and we will speak more at another time about what is truly harmony for you.

Q: (M) I find, if I move into this base of allowing or embracing, that fear comes up, like a fear of not being in charge of a situation, which makes it somehow hard to contain.

P'taah: Indeed. Do you know why it is? It is called fear of annihilation. In your culture you are taught that you must always be in control. If you are not in control, somebody else will be and then you will die. Now, even that would be alright, beloved, if you would understand that truly there is no death. Beloved - it is alright. Allow, allow the fear. Do you know, it is quite permissible for you to say to someone: 'I am really frightened, because I feel I am losing control.' It is to know that surrender, beloved, is only surrender to SELF. It is that allowance that will create for you light, will engender love. Do not judge yourself for being afraid. Everybody is, you know? It is alright. You know, beloved ones, it is really quite amusing, when you think about it: All of you think that everybody else has the answer. All of you think everybody else knows and that you are the only one who does not and that you are the only one who will miss the boat. There are those who are so very learned, there are those who are very good at speaking spiritual platitudes - and that is alright. But it is for you to remind yourselves that for all there is a wise platitude that says: every soul is really at the beginning. Now, when you understand cyclic harmony, then you will understand that there is truly no beginning and no end. We do hear some say: 'That is a very old soul.' Well, you are all very old souls, very ancient. Question?

Q: (F) P'taah, I find time highly accelerated for me, I do not seem to be able to keep up. I go to sleep in the evening and the next moment I go to sleep again. Time seems to be flying.

P'taah: It is very simple: stop it.

Q: What do you mean? Just stop doing what you are doing?

P'taah: You can change time, beloved. You are a very powerful entity. You may change time. Speak forth and you may say: 'It is all going too fast. I desire more time' and so it will be. You may *take* time. However, beloved one, we will admit to you, that time is warp time, hm? You are not the only one who is finding this *[to be so]*. It is alright, but you may change it, if you wish.

Q: How?

P'taah: *Know* that you can and take time.

Very well, we will make a break, that you may refresh your bodies and we will return, when you have formulated more wondrous questions for us. We would ask that you be silent for two minutes. I thank you. You know, dear ones, it is such a joy to be with you all. Very well.

(After the recess:)

P'taah: And so dear ones, refreshed in the body? We shall continue with questions.

Q: (F) When we are born, do we bring with us a gift or talent that helps us sustain or support us in our life?

P'taah: That is a very good question, dear one. Much is made of the drama and trauma that is brought forth with each lifetime. In truth, one of the greatest gifts you bring forth at birth is called soul integrity. That is the will, the desire at soul level to come forth into life to flower, to have within the readiness to allow all that may occur. I speak in the divine sense, do you understand? Now, certainly when you are thinking of lifetimes, when you are looking back so to speak and you are thinking in terms of lifetimes after lifetimes - you always think of them chronologically. *The truth is that past lives and future lives are really occurring at the same time.* So, in this fashion there is always the greatest potential and there is what is not being explored in other lives, you understand? As you are developing your interest in what is occurring in your other lifetimes, many people are using various methods, called regressions, to come into an understanding of what they have been up to in those other lives, yes? What many do not understand is that you may also explore the 'future' lifetimes, because outside of the localized time that is this space-time continuum, it is already in motion. Now, many in this exploration are getting pictures, which they feel to be past lives, but in fact very often they are not past lives at all, they are what you would term future lives. We will also not discount that which are *probable* lifetimes. Now, this is where it becomes tricky. You are multidimensional energy and there are not only past lives and future lives, there are also the dimensions of frequency. We have said to you that in truth you are already ascended masters, that you are everything. The soul energy is not confined to this planet or this time - but there is also that which are the probable realities of every lifetime. So we are truly speaking of an infinite number of lives. Now, if you were to become very involved in all of those other lifetimes, indeed there would be no time to be involved in this one. And truly, it is not important, because each life is like an encapsulation. If you can truly be involved in this life, if you can truly be involved in each now moment to live as fully, as richly as possible, then you would certainly have no desire to escape to the thought of other lifetimes. But it is indeed to know that the greatest possession you have is your SELF, because in that SELF there is all honour, there is soul integrity. There is limitless possibility that you may flower into your own potential.

(The importance of the following dialogue warrants an additional comment: A gentleman is concerned with traditional spiritual teachings, which project a personification of the Source in contradiction to the understanding of the Source as being within, as being the I AM. P'taah clarifies the seeming contradiction by pointing to the humanity of our present time as having a greater capacity to comprehend a more expanded understanding, thus viewing the teachings of the past as being perfectly in order for that time.)

Q: Beloved P'taah, we as individual extensions of the Infinite are a part of the Source, we are all One and we should look to the God within. Yet the ascended masters always give great reverence to the Father, the Creator, which seems to be without. HE also seems to be a beloved brother as well as the boss, for example the Lord's prayer: 'Our Father..', could you, please, comment on this?

P'taah: Indeed. That which have been ascended masters, that which have been your role models in your history have been speaking each time to the collective consciousness of humanity. The collective consciousness of humanity at this [present] time is in waxing, is in a state of wondrous change. At this time humanity may embrace a grander idea construct. When we say that it is worthy of thought, that the Source is more than a personalization, then understand that it is almost like a quantum leap for humanity. Because, thus far in your cultural history, it has been that humanity has been in need of an authoritative figure and so it is an easy concept for humanity to think of a great Mother or a great Father figure. You understand? And indeed, we speak forth about the God/Goddess and we speak forth about the great entity,   your Earth, as the great Goddess, but indeed it is a grand concept for humanity to understand the Source as being far grander as merely imagining an authoritative figure. Do you understand? We are certainly not invalidating the words of ascended masters, dear one. We are suggesting merely that as you extend conscious thought, so you are opening the 'radio antennae' to the greater of who you are. Does that answer your question?

Q: Yes - I have to think about it.

P'taah: You do not need to, you know. You already know it. Indeed, it is wondrous to see you again. And how is it in the grand Outback? *(P'taah refers to Queensland's interior, where the man spends his professional life.)*

Q: It could not be better.

P'taah: It is grand, indeed. *(Another gentleman addresses P'taah.)*

Q: Since I do not know what is truly best for me, how can I continue to desire? Does it not interfere with the necessity of surrender?

P'taah: It is not so, that you do not know, dear one. It is just that you do not know with your logical intellect. As you exercise surrender, so you will come into what is called, intuition. Your logical mind has been a wonderful tool for survival, but now, you see, *survival of your species depends on intuition, on nurturing surrender.* It is not necessary to know, only to feel. *(P'taah speaks now with great tenderness).* To imagine, to allow love to truly blossom within you. It is called a leap of faith, because, beloved, who you are is grand indeed. Who you are knows the divinity, who you are is one of honour, of integrity, full of creative joy. Who you are is waiting for flowering. Do you think, beloved, that the flowers worry about how their roots are placed, do you think they worry about if it were to rain or not? They do not. That is called integrity - to know. Every cell in your body knows and so do you. You have just forgotten that you know. So it is to trust who you are, to allow the knowing to come through. We will speak to you more and you will find that you will unfold into the knowing. It is no 'big deal'. There is no difficulty, there is no struggle. Struggle belongs with physical survival. Surrender - belongs to 'survival' of all that is wondrous in the spirituality of humanity. Questions?

Q: (F) You spoke about understanding through resonance with divinity. Does that mean we need not learn anything because we can know by tuning into the Source?

P'taah: Indeed. That is called the shortest answer on record. *(The audience is amused).* That is exactly how it is. Within the cyclic changes of humanity, that which is regarded as education, so necessary for your young, will change. Children born now and over recent years are already in a different knowing from those who were born more than ten years ago, indeed, perhaps a little more. Children born now are born of dolphin consciousness, are born of star seeding and have more intrinsic knowledge; they will create great changes because there are many things within your present education system that they do not need. There are many things that are being created genetically to change humanity, all being brought forth to assist in what is really a quantum leap in consciousness.

Q: Could you comment on inherited illnesses and their connection with parents and grandparents?

P'taah: Dis-ease of the body is emotional disease. We have spoken of this. *Belief structures about illness create illness.* Where one may have an inherited disease, it is not written in stone, it may be changed. We have spoken of the morphogenic resonance of the cellular structure of the body and indeed the morphogenic resonance of the family, the culture of the country etcetera. When a family believes that a disease is inherited, so it is indeed. That which is the genetic program of an inherited illness is a minute proportion, and we are speaking merely of physicality here. What you believe *IS*. We have spoken about medicine; you believe it will fix you and it does. You believe you are a victim of disease and you are. Why do you think it is so prevalent that if one person of a family has a disease, that the child will have it?

Q: I thought possibly as co-creators we chose our parents - we chose a body which reflects our own consciousness. Diseases are a result of our consciousness, therefore we resonate with the morphogenic resonance of people who have the same symptoms.

P'taah: Very good. And yet, it can be changed. Where there is an impairment, for instance, when a person is born physically impaired, it is certainly chosen and it is certainly co-created when members of a family come together with physical impairments. It is a wondrous opportunity for people to understand that the body is not a limitation.

(Another lady extends the subject of illness in regard to animals:)

Q: P'taah, can you explain disease in animals? Is it their karma or are they reflecting our consciousness? For example: One of my chickens is developing a cataract in its eye. I do not understand how to relate this.

P'taah: When you are speaking about domestic creatures - those emotionally tied to you, as with your cat and dog and horse, where you have a very close emotional bond, then very often the creature will display physical symptoms not only for you to reflect on, but as an idea of its own relationship with you. Where there is no emotional bond, we would say that very often what the creature is creating is for its own experience. However, dear one, as it is your chicken which is creating a problem with the eye, then perhaps you may look at how it is you view your chicken.

Q: Oh, I see.

(Delighted with the double implication of that answer, the audience bursts into approving laughter.)

P'taah: Very good.

Q: (F) I was wondering, do crystals help to lift one's level of vibration and awareness?

P'taah: Crystals are tools, no more - in this fashion. Your Earth is a crystal. Humanity is also crystalline. But you are speaking of the crystals which you hold and use as objects of power. Indeed, they are a tool and they can be used as magnifiers. They also serve to store information, but you truly need none of it for your own raising of consciousness. But if it pleases you to use crystals, then it is valid.

Q: (F) It is possibly easier to understand people choosing their experiences, but how is it with animals? Do animals choose experiences like we do? We see animals maltreated. Do they choose to go to places where they are maltreated? How does that work?

P'taah: Dear one, all - all is co-creation. Animals of domestic nature are truly to reflect emotion to humanity. It is also a great opportunity for learning. *Whale and dolphin - they are the same soul energy as humanity.* They have chosen to create a horror story for those humans who are aware. It was their conscious decision. The second density creatures do not choose consciously, any more than most of humanity consciously chooses. As humanity comes more and more into its own knowing and as its consciousness expands, so does the consciousness of all things upon the planet. It is a resonance. As the vibrational frequency rises, it rises everywhere. So that when humanity steps forth into fourth dimension, so does every creature, so does every blade of grass.

Q: You said all of humanity?

P'taah: Well it is not like all, truly.

Q: Do all do it at the same time?

P'taah: Those of you who choose to.

Q: But if some choose to do it sooner than others, how is it then with animals? Do you understand what I...?

P'taah: But of course, I understand exactly, beloved.

Q: For instance my cat. It seems to be .. gosh..

(The lady finds it difficult to express her thought and P'taah tells her kindly that he will attend to someone else's question first, this will give her time to clarify her thoughts.)

Q: (F) P'taah, do animals have the choice where to exist?

P'taah: They do. *(After a brief pause P'taah attends to the previous lady once more.)*

Now, what you are saying is this: When the change from third to fourth density occurs, everybody is worried - and this is a slight digression - about those who will not go forth into fourth density, hm? So, now you are worried that some animals will not go forth also. That is alright, beloved. Indeed, it is very good - that you worry. - *(Pausing)* - I am teasing you. We are speaking truly about compassion.

Q: Sure.

P'taah: As humanity gathers its energetic forces for the grand shift, it is not ... You know we find this very difficult to speak forth in words, which make sense to you within the boxes of your understanding. *There will be certain humans who choose to stay in this dimension and there will certainly be, in their creation of this reality, animals.* However, at another level of consciousness all, *all* animals will go forth in that shift.

Q: Is it so that the soul of an animal will go forth, but will leave something like a holograph behind here?

P'taah: It is not quite as simplistic as that. Soul energy of animals is, in a way, fragmented. If the soul energy of third density beings - humans - decides that it would desire to experience that which is animal - now, this is not done through third density consciousness - it is not that the whole of the soul energy goes into the creature, but that a fragment of the soul energy may do so for the experience. *People do not turn into creatures.* The soul energy may also decide to experience itself in crystal formation, as a rock, as flora, do you understand? At a certain vibratory frequency the soul can experience itself in any density.

Dear ones, you are all looking very tired.

(A gentleman has one more question:)

Q: Beloved P'taah, one more question: The collective [human] consciousness of the planet Earth seemed to be quite destructive ten years ago. Is it improving at a better rate now than what it was ten years ago? Are we coming closer to the quantum leap?

P'taah: In the fashion of cyclic changes you are certainly coming closer. And certainly it is closer, because in those ten years there have been many grand people who have decided for love and life, instead of death and destruction. Does that answer your question, dear one?

Q: Yes.

P'taah: Beloved children, sufficient unto the time. We are delighted to see you. *(Addressing the host of the session:)* Our thanks, dear one.

Q: My thanks.

P'taah: *(Addressing the hostess:)* Indeed, our thanks, beloved woman. *(Then to the audience:)*

Indeed, it is always joy to my heart to be with you - and in your expansion and in your desire to truly know your own expression of divinity, so indeed are you bringing forth expansion for all energies on all galaxies and we are truly in awe of your courage. It is truly a delight to be with you.

Good evening.

Chapter 5

FIFTH TRANSMISSION.
Date: 25th of September, 1991.

P'taah: Dear ones, good evening.

(P'taah looks at each person individually, acquainting himself with each person's energy and - or - state of mind.)

Well dear ones, pleasure indeed to be with you. So, the adventure continues. Now, this evening we shall continue to speak more to you about transmutation. It is very important that you understand exactly how and what it is. *(Addressing a gent:)* Already questions, beloved? Hold it for a while, for I wish you to listen not only with this, *(pointing to the head)* but with this *(pointing to the heart).*

So, it is called 'hold the phone' - is that not correct? See, *(jesting)* I know all of these things. Now, we will recapitulate: Transmutation - the ultimate alchemy of the soul. To transmute agony to ecstasy, and how do you do it? Very easy steps, beloved ones, very easy: take responsibility to know that, indeed, you create your own reality absolutely. All, all within your universe, within your personal universe, is a co-creation. Step two: judgment - to align the judgment. Now, how do you align judgment? It is simply to acknowledge and accept that the fact that you are judgmental is indeed valid. It is who you are. It is also divine expression. Do you understand? It is divine expression because it IS, because it exists, and the only way that you may create the changes is by the embracement, by the acceptance and acknowledgment. That creates the alignment - by acknowledgment. To say 'I am judgmental, I am judging myself, I am judging the situation - I am judging other persons involved - in whatever painful situation it is. I have created it, that I may learn the lesson'.

In this way all comes into harmony. To know that pain, which you also judge so harshly, is what encompasses the pearl of wisdom contained within.

Now, as we have said before: it is very difficult for you to align the judgment, when what you really wish to do is to commit murder. We are jesting. Well, we are not really jesting, because indeed very often that is exactly what you wish to do. *Take responsibility, align the judgment and feel the feeling.* We will remind you again: *Pain is not feeling - pain is resistance to feeling.* And as all comes into alignment, so indeed the energy, which is held within your belly by resistance, is released and goes from the solar plexus to the heart. It is truly that simple. It is not something that you do. It is to be allowed. *In the allowance the transmutation occurs.* At this stage I would ask you if you have any questions about this process. — *(As nobody seems to have any questions, P'taah responds with a smile:)* Very good, you all will go to the top of the class.

Now, in the simplicity of this operation it is very easy for you to be bogged down because you try so hard to escape the pain. Pain is only the resistance. It is not to escape from [pain] - it is to encompass, to embrace, so that you may *feel.* Beloved ones, you have forgotten how to feel. You are so afraid that you will die. It is called invulnerability. When you feel joy, when you are about to experience something which will create happiness within you, but you judge it to be unpleasant, then immediately the clamps come on. *Take responsibility - align the judgment - feel the feeling.* That is all.

The changes which are to come to your planet will come, whether or not you are expecting them; whether or not you believe it to be [so], and all of this, dear ones, is truly that you may be in harmonic circumstances to reap the reward. Because when we speak of harmony, that you may all be harmonious within, we are saying that *all* things will be in harmony. That is the New Age. That is what is to occur. All of you so desperate for enlightenment are in truth quickened by the whole of your universe gearing itself to the change. That is why we are here; and we are not speaking in the royal 'we'. We are speaking of all of the entities and of all of the beings seen and unseen, to assist the transition of the Earth; the transition of your planet, the entity, your Goddess the Earth and every being associated with your planet. We have spoken before about the importance of Earth humanity coming into their spiritual fullness that they may be in harmony with their planet, that they may be in harmony with the new technologies to come - all which will come to fruition much more quickly than most of you understand. You will not come into your spiritual adulthood until you can understand that you may be whole. It is not a 'pipe dream', dear ones. To come into this fullness, to come into wholeness, it is necessary that you can take the pieces of you, the facets of who you are, that you judge so very harshly, and embrace them into SELF. To take situations which are painful, whether it is present pain or whether it is pain of something long gone in your past - truly, every time you are in pain, you are only repeating a situation. It does not matter what the story line is, it does not matter *how* your soul has brought forth a situation for you to examine yourself and to embrace. This is why, when we are speaking with you, we always bring it back to SELF. Because *until you can enjoy all facets of who you are, you will not be whole.* When you are whole, dear ones, it is *coming home.*

Your planet is coming home. And earnestly we desire that each and every one of you may join all of the beings who are waiting to welcome you.

Now, in your lives there are very many areas with which you are not in harmony, discordant facets, because of your beliefs. We shall, in the next evenings together, discuss with you some of the belief structures that you hold, that you do not really understand that you have. Some of these beliefs are keeping you in chains, because all of it comes back to the judgment you have of who you are. Now, we would say: General questions to start with.

Q: (F) Hello, P'taah. About these changes, does it mean that even though the Earth is going to change, if we do not change ourselves, we will not come into the change?

P'taah: Now, we have heard this many times, when entities are speaking about the changes. There occurs a bottom-line terror that you will not make the grade, that you will not be enough, that you will be left behind, and if you are convinced that you will not be left behind, there is the terror that a loved one will be. That is very valid but it will not occur like this. In a fashion we could say that nobody will be left behind, and really, it is not 'left behind', you understand? We would say to you, there is nothing to do, and nobody will be lost in the process. When the time for the transition comes forth, nobody who desires the transition will not feel it, indeed the whole Earth will resonate to the change. *However, the moment before the change occurs, there will be situations upon the planet which will make people waver.* People will doubt themselves and very often, when people are faced with fear, [they] react in quite extraordinary ways. War is a very good example of fear, indeed? And if people believe that they are facing annihilation, if they believe they will starve to death or that nature has gone crazy and that they are going to die, then they will react in extraordinary fashions. So it is for you to know that the coming changes are really something wondrous, because they herald the wonder of the transition. Now, when we speak of the Earth changes, of the changes within politics, within economics, within technology, concerning every facet of life; there are always the harbingers of doom, 'the end of the world is nigh'. Indeed? We wish that you will remember that whenever you see something which seems to be absolutely cataclysmic, it is to know that you are indeed one step closer. Does that answer your question, beloved?

Q: Yes.

P'taah: Indeed.

Q: It seems to me that instead of thinking of things to come, that we can already embrace death and disaster in our own time now. We can already work through them, so when they come they are already done for us.

P'taah: Indeed. That is why we are here, so that people may begin to reflect, to have the knowing, not merely as an intellectual exercise, but as a knowing within the breast and certainly within your own time. Do you understand?

Q: (F) Yes. Just recently in conjunction with your teachings I have been contemplating polarities and their embrace and I had a few opportunities. Pain, headache, fear of death. I felt that it is

*alright to die - welcome it. Is that part of the process? To say death
is death and contrary to our belief system it is ok?*

P'taah: There is no such thing as death. When you become very
tired of wearing one colour every day of your life, it is a joy to take
off the suit of clothing, hm? Your body is, in a manner of speaking,
a suit of clothing. When you take off your clothing, you do not cease
to be the wondrous jewel you are. There is no such thing as death in
truth. Fear of death is quite extraordinary. It *[death]* is merely a new
adventure in consciousness and when you truly have no fear of death
anymore, then you will know, beloved, that you have taken a giant
step forward. Because with the absence of fear, you know it is no
longer an intellectual exercise, but a knowing deep within you. It is
indeed a transition, of course, as birth is a transition from being a
foetus to a breathing organism and all that goes with it. But the
consciousness goes forth and on and on, even if you do forget.

(P'taah turns to a gent.) And now, dear one?

**Q: Beloved P'taah: I recently read that it could be very beneficial
to mentally give permission to energies surrounding us, to enter our
being to enable a better understanding of our true reality. Could
you, please, comment on that and possibly instruct us?**

P'taah: But indeed it is beneficial that you give yourself permission,
because that which are other energies and that which is your soul
energy is always available to you. You see, humanity believes that it
is alone and even when they know intellectually, from all the
wonderful reading, that they are surrounded by energies - entities -
it is about the knowing within the breast. So, when you give yourself
permission - and truly all it is, is to relax - ask and it shall be given
forth, dear one. Allow for whatever is to occur - whatever. You know,
it is also quite a paradox, because very often you will say: 'I have a
very knotty problem within my head' and you say 'oh, universe, send
forth the knowing, send forth the information to solve this knotty
problem' and you are all waiting for celestial voices. Well, very often
it is not in the form of celestial voices at all. It can be something quite
mundane which occurs within your day to day life, and if you are
aware you will say: 'Aha, that is the answer to my question'. Do you

understand? The moment that you have expectation on how the knowing may occur and when, you are immediately shutting down the possibilities. So, it is to put forth the permission to ask whatever may be set forth, then it is to *be watchful.* You know, your soul is often very tricky. So do not rely on 'angels with gilded wings'. *(At this very moment the house cat is crossing the room and P'taah uses her as an example:)* You may, [for example] rely on the cat coming forth to teach you a lesson. — Indeed?

Q: (F) When you say that pain is not a feeling and you should just allow the deeper feeling behind it, so when there is pain and I say now I allow whatever is there, do I first feel the pain and wait for whatever feeling comes?

P'taah: Now we are getting down to the nitty-gritty, dear one and thank you for the question, because indeed it may be very perplexing and we wish that every body be very clear on this. Imagine this now to be real, where somebody you love very much has died and - even if you are in the knowing about death being only an illusion and transition - if the person you have lost is somebody you love, then the loss of physical proximity to that person brings forth grief - pain. Pain, indeed, of the heart. And it is to know that this pain is not only the grief of separation, but it is the pain of eons and eons of separation. It is not to judge the pain. It is not to think that, because you are in pain you will not make the grade. It is to acknowledge the pain indeed and to know that you called it forth - growing, accumulating - it is to accept it. Acknowledge it, embrace it in the knowing that it is a valid, divine expression; that it is alright, do you understand? If you are using rational thought to suppress the pain, you push it down and if you are trying to escape the pain per se, that which is transmutation will not occur. It is like being at the seaside on a very hot day and looking at the water, but until you dive in you will never know how it feels, how refreshing, cooling - how it makes the nerve endings of the body tingle and come alive. And so it is with pain. If you will embrace pain without judgment, the resultant ecstasy is *more* than the swim is on the hot day. Then you will truly know what feeling is.

Does everybody understand when we say that pain is resistance to feeling? *(The audience is affirming this.)* You are very clear.

Q: (M) P'taah, regarding the last session: In parting you expressed admiration for our courage. Did you refer to the courage needed to dismantle existing belief structures, or what did you specifically mean?

P'taah: It was not so specific, but indeed manifold, because it is quite courageous, you know, to choose this lifetime, any lifetime indeed within this field of reality; but also the courage of the individual that you are, to go forth in an act of faith, because indeed, dear ones, that is what it is. It is an act of faith because there is no guarantee, and we understand that humanity does indeed love a guarantee or your money back. It is all based on a desire to be more. It is also in the harmonics of your time. The very air is quickened with expectancy. But there are many who are perplexed but will not look because they are afraid. Because, dear ones, what you are doing does not exactly fit in with your normal social structure. There is much judgment, even for our woman, who would say that if her family would know - not her entire family - what she is up to, they would bring out the straight-jacket. I am sure, dear ones, that you all know somebody who would bring out the straight-jacket, if they knew what you were up to.

Q: But should it not be expected that humanity adjusts and opens more and more to spiritual education?

P'taah: But of course, dear one, this is what is occurring. And you know, sometimes people become despondent and they say: 'Well, if all of this wondrousness is going to happen very soon, how come everything is in such a mess?' 'There are still the wars, there is still the starvation, there are still the murders, there is still discordancy everywhere.' But you see, it will become more and more like this - *and then it will change.*

Q: So, in other words there is a split in consciousness occurring and does that split have to go rock-bottom or...?

P'taah: Dear one, it is not truly a split in consciousness. It is merely that some would like to stay to maintain the status quo and they are fighting an uphill battle and that will become more and more so especially when the *Earth changes* become more and more apparent. For then there will be also rampant the feeling of powerlessness. It is very hard to hold on to your financial structures and your bureaucratic power, when your country is falling to pieces. *That will not occur in this country.* You will not know the devastation that will be in some *[countries].* So, you have all chosen extremely well, hm? And we will also say this, because there has been much talk about the devastation of the planet and much fear engendered; well, in a way it has been a good thing, my dear ones, because it is as my woman would say: 'Get them off their asses.' We will say this: as you come more and more into your own knowing and more in tune with what is occurring - and we are not even saying that you are consciously aware of this - that if something of devastation is to occur, you simply will not be there. What we are saying is: Do not cancel your holiday trip, do not cancel your day-to-day living in case of disaster. *It is to know that you live in a safe universe,* beloved ones and you do - you do - you create it. When you are in love with who you are and when you are showing that love forth to your planet, how can you be anything but safe? *Not to be safe is to be in fear.* When you are in fear what do you draw to you? Disaster - not safe, indeed!

Q: (F) What you say, I have experienced it - about being in a safe universe, about death being a transition - I have experienced that space, and yet there is still part of me which is afraid. There is still part of me which fears annihilation, which fears death, which feels that I am not worthy. I just look at my life - there are beautiful things happening, and yet...Yesterday I experienced for a little while an absolute Self-loathing - it is there, so...?

P'taah: Beloved, why would you be different to anybody else? I am teasing you.

Q: Yes, I know. It is simply that I ask you to comment on those things. We know these things and yet within there is still this other..

P'taah: Of course there is. So first things first. What is the answer, dear one? Indeed - surrender. The feeling of Self-loathing, the terror of annihilation is called the base line of humanity. For many the terror of surrender is greater than their fear of death. So, tell me, what did you do when you had your moment of Self-loathing?

Q: I thought about it a great deal. It went around and around in my head a great deal and eventually I recognized that it too is a valid and a divine aspect of who I am and then it went.

P'taah: How extraordinary, beloved. It simply faded away, eh? And yet, you know, if it truly was a terror of the moment, you could transmute it. Obviously, beloved, it did not occur to you at the moment.

Q: I think my head was...

P'taah: ..in the way. We would say to you all: When you are feeling great pain, when you feel confusion, when you feel fear - there is a very simple answer: Take off your head and put it under your arm, it will be of much more use there. *(P'taah is facing an amused audience.)* Because, beloved ones, you know, truly your life is about feeling, emotion, creating feeling. When you can allow the channels to be open to the feeling, then there is no head there. You do not feel here *(P'taah points at the head)*, you feel here *(points at the solar plexus.)* So, take responsibility, align the judgment, take off your head and feel the feeling. *(Laughter)* We will insert that in the new rules, eh?

Q: (F) It seems to me that we have brought this thought into spirituality, that we are not supposed to have these things happen to us anymore. So we always feel very guilty when we are in pain. But it seems to me now that we can throw this thought away and as you say, it is alright to cry, it is alright to feel sad.

P'taah: Beloved, when you are coming across very spiritual persons who do not feel pain, who are very - what is the word now - sanctimonious, hm - then you may know that they have what is called a good front, unless you can actually watch them ascend before your very eyes.

Q: Also, we seem to give too little credit to the joys in our lives. For example: When we have an argument with a friend, then that is a reflection. So we feel so guilty about that bad reflection, but you are also a reflection - we have so many beautiful reflections that we do not validate. We seem to notice those more, which we do invalidate.

P'taah: Indeed - and of course, you know that 'spirituality is so serious'. And when you are spiritual, you are so *good*. Every day of your life you have a myriad reflections. Some of them drive you crazy, hm? We are speaking of 'people' reflections. In fact, we often wonder what it would be like if you had no nature to reflect to you your own beauty, because it seems to me that very often your 'people' reflections are very sombre and not very much fun - very often. So, perhaps that is why you have created nature, so that you may truly understand your own grandeur, your own beauty, your own awesome strength and power. We have said before, spirituality is not about being good. It is about *Being - human Being*, every moment, however it is. In all its joy, in all its pain and sorrow it is to be in the moment and know how exquisite it is. Even the rage and the anger - whatever it is - you are alive, you have chosen it. It is a tempestuous path to enlightenment. Be grateful for it. It would be very boring for you if you did not have it. You are addicted to it all. You love it. If you did not, it would not be and that is the wondrous and very humorous dichotomy of it all.

Now, dear ones, we shall take a break for a moment, so remember your questions and we will come back to you very soon and we would like to have more questions. It is by the searching, by verbalizing that you may reflect indeed and wonder indeed, that as you formulate your questions, you already know the answer. So enjoy your refreshments.

(After the break.)

P'taah: So, dear ones, you have prepared many questions?

Q: Is fear a true feeling, or is it the absence of feeling?

P'taah: This is a very good question. Fear is a polarity. Now, the humanity indeed is truly sovereign. Free dominion. Free-dom. Always, there may be choice. You may choose which polarity, fear or love. So when you are in the grip of fear, you may catch yourself and you may say: 'Now, which do I truly want?' Very often, you will find that, upon examination, fear is rooted in your belief about yourself and about reality *as you perceive it.* When you change the belief structure and change your belief about who you are, then there is no fear. However, dear ones, let us not forget that again: fear is valid, it is an aspect of who you all are. It is, like everything else which is not love, transmutable. But you do not transmute fear by trying to suppress it, by trying to push it away, or by trying to ignore it, because it is an invalidation; and we will repeat, what you invalidate, indeed you empower. Fear, polarity of love, and whatever it is you are feeling, you are projecting into the universe. So, *whatever it is you are fearing, you will draw to you.* It would behove you well to be conscious of what you are thinking. Your thoughts are power. Your thoughts are electromagnetic energy. *Your thoughts create your reality.* Your belief structures are the idea construct of your universe, and you, dear ones, are the central sun of your universe. Always, always, it is for you to examine what it is that you do believe. We have spoken forth many times that your belief structures are very often like prison bars. They are boxes that you inhabit within your consciousness and now is the time to collapse the walls of the boxes to move into an expanded consciousness, an expanded awareness. All you need to be aware of is to put forth the desire and you will find that you will bring to you situations for the learning, for the understanding of what you believe, day by day. What do you believe about who you are, hm? Utmost simplicity, this. What do you believe about yourself? And it is not to negate any part of it [the belief structures], it is merely to accept and acknowledge that that is who you are. That is how you grow. So simple, dear ones, so very simple.

Q: (F) An entity told one of my friends that alcohol is the worst thing for us. Can you tell us your views on that?

P'taah: I am very fond of it. Alcohol is the worst thing? That indeed is a fine judgment, eh? And if you believe it is, then certainly, it is. As we have said, dear one, *when you are aligned, you may drink rat poison and it will be alright.* We are not suggesting that you employ rat poison. Now, let us speak a moment, again, about the 'dos and the don'ts'. Do not eat meat. Do not eat vegetables grown in water *(hydroponically grown)* instead of [grown in the] earth. Do not drink alcohol. Do not... what are some of the other wondrous ones? Do not smoke cigarettes. Yes, we have been speaking to our woman about this. Dear ones, do everything, if it pleases you. Balance is, perhaps, a watch word. There are many ways to the path, and who is to judge that one may find enlightenment at the bottom of a glass of the ascended grape? Now, we are not suggesting over-indulgence in anything. If you are living on a diet of tomatoes, you would become, probably, physically debilitated. However, we will say, to consider your body to be an extension of your light body, that which is your soul. It is precious, it is beautiful. If you do not treat your body with honour and respect, how can you treat anything else with honour and respect, hm? If you are in the belief that there is anything which is really bad for you physically - we hear this very often, it is very bad for you - it is reinforced all the time, all the time. We will say this - nobody dies from the effect of alcohol. Nobody dies from the effect of tobacco. You die from broken heart. It is just that you make it very simple to exhibit the diseasement within. When you are engaged ingesting chemicals which do not give the body joy, then it would be wise to think about it. We will not say alcohol is the worst thing you may take. When you are in such rigid dogmatism, you are not allowing a flow of information, and always, always, your body tells you how you are. It is a mirror. So you may indeed enjoy your glass of alcohol. *I do indeed, very much*, much to the disgust of our woman, sometimes. But it is to know, beloved, that if you are truly in love with who you are, you are not affected. Questions?

Q: (F) P'taah, I am experiencing a lot of anger at my work place at the moment. Anger at myself, anger at others, and I am observing a lot of anger, and it is valid. But what is anger? Sometimes it just dissipates, and at other times, it stays with me for a very long time.

P'taah: Now, what is it, do you think, that lies beneath anger?

Q: Judgment? Negative judgment?

P'taah: What about fear? What about hurt, and pain?

Q: I wasn't thinking of that.

P'taah: Now you may think of it. And you are quite correct, beloved. Anger, quite glorious expression, perfectly valid, else it would not be, and when you are in it, enjoy it. Have a wonderful rage. When you are truly angry, and very fiery, you are certainly living in the now moment, especially when you throw things. *(Laughter.)* We have said this: it is alright to express anger, but then it is also to say: *Why?* When you know why, and when you are in touch with the pain, then you may transmute it. Take responsibility, align the judgment, feel the feeling. Oops, I forgot - take off your head. *(Laughter.)* Do you understand?

Q: (M) It is really a joy being in your presence, and in the last weeks it seems that you are with me all week and making my life really intense, and bringing me to new dimensions, somehow, so I wonder is this really happening, and what are you doing?

P'taah: I see, I get all the blame, eh? This is not new for me. I am strong enough to take it. Beloved one, you are coming into new dimensions of who you are. Is it not wondrous? But it is you who does it, it is not I, because you are the creator of your reality. Now, you may certainly call me forth to share it with you, but it is you who creates this wondrousness. It is you who raises the harmonic. It is not I who is doing this. That is the capability of all of humanity, not only for the energy exchange with that which is I, but indeed, the energy exchange with all of your brothers and sisters out there in the universe. And as you raise the harmonics of your own vibratory frequency, so it is that you will draw to you wondrous energy from all of humanity about you in this dimension. Because you see, beloved, what occurs is, that as the frequency is raised, so you become irresistible. Is that not wonderful? Indeed. When you return to your home this evening, it is for you to look into the mirror and know how wonderful you are and give thanks to who you are.

Q: (F) P'taah, some days I wake up and I have all the highest intentions to have a most wonderful day, but all the odds seem to be against me, like a handicap race. As well I noticed that when I come into my pre-menstrual time it seems especially difficult. It is as if the odds are insurmountable. Can you comment on this?

P'taah: Now, there are two things here. One is the shift of balance within the cycles. One is the belief that it should be so and many women believe it [to be] so. It has been a great propaganda, what is called PMT, hm? Do you know, not so long ago women did not understand that they should become misaligned of the body. Now it is certainly the belief structure of women which has created, in fact, a valid discord, do you understand? It is valid - it is real, but it has been created for a grand purpose. Now, the days when it all gets too much, where somebody puts a stumbling block in front of you, you know, then you may say to yourself: 'Well, here is the stumbling block, it is going to be one shit of a day.' *(Great amusement in the audience.)* Then relax and enjoy it. Having a shit of a day is quite valid, you know? *(More laughter.)* One of the important things we feel is that you should truly have a sense of humour about it all and when you can laugh at the stumbling blocks and say: 'Well, this is a shit of a day', you will find it will all change. Because you know, *when you are in laughter from the belly, you are in true alignment,* because when you are having a grand laugh, there is no past and there is no future, there is simply *IS.* It is called the joy of the moment, it is called fullness, resonance with the universe. That is why we are not too much on the 'big', on the what is called 'heavy' stuff. I do not need to do it, because you are all very busy doing it all the time. So we are not going to be serious. Do not judge yourself, beloved woman. As we have said, humanity is living life learning from every experience. When you are becoming an ascending Goddess, you will not have a shit of a day. It is alright - it is called surrender. *(And quietly and tenderly he says to the lady:)* I love you.

Q: (F) P'taah, I would like to ask a question about dreams. I have fantastic dreams most of the time and recently I had one about a very, very large landslide and on the night news I heard there had been one in China. I was wondering if this vivid dream was actual reality, that I actually witnessed this or was it just my night wandering?

P'taah: On this specific occasion you were certainly tapping into that which was a physical occurrence, but all of you do it. Very often in your dreams you will dream of something which is occurring on the planet, but because you do not have notification, you do not know, hm? Now, of course, there are things in your dreamtime which are what you would call prophetic. But you see, what you are not understanding is that there is no separation between all of humanity on the face of your planet. There is no separation in truth from everything occurring on all dimensions of reality that you consciously tap into. To remember it all is perfectly natural. The thing which is not natural is your concept of separation. Now, we will speak again briefly to remind you about your dreamstate. Your dreamstate, like everything else, is multidimensional. Very often the information being processed is not acceptable to your conscious mind and so you create pictures that your consciousness will understand. Very often your journeying in your dreamtime is symbolized by flying, by being in the ocean - very often you are meeting with your friends. Sometimes they are friends you do not know in this lifetime, in this conscious reality. It is multidimensional. Humanity, no matter how often it hears it, insists that they are separate. You are not. All of you, truly understand, that there is no separation, that there is no separation between you, your planet, every single human, every leaf, every flower, every rock and creature - you are all woven into the one tapestry. You are not separate from your soul - you are not separate from the divine Source, you are not separate from any facet of who you are. Who you are is wondrous indeed and as you come more and more into the acceptance of the wonder that you are, so more and more in your consciousness you will be understanding in its fullness the multidimensional aspects of Self. Until you acknowledge who you are, there will always be the pain of separation. And when you can allow transmutation to occur, when you experience the ecstasy of Oneness with the whole universe, so little by little the separation will become less and less. *It may only occur through the acceptance and acknowledgment and the love for SELF.* It does not occur by hiding in fear.

Who you are in your grandeur is awesome. What you are truly, beloved ones, is one facet of every jewel of your universe. Every facet of you makes up that facet. There is nothing about you, no-thing, which is unacceptable to the divine Source. You do not have to do anything - *it is simply to be who you are* and in the being, and in the acknowledgment and in the embrace of every facet you create the change you so fervently desire. But you cannot *do* it - you can only *be* it. And each of you is a part unfolding - each of you with your face turned to what is the Divine Light - and each of you is coming into the understanding that you are truly divine expression. *What you truly are is God smelling the rose.*

When you leave this place this evening, look into your heavens and see the glory of the Moon and know that you are also in that awesome beauty. She knows how beautiful she is. It gives her great delight and so it is that the contemplation of your own beauty - no matter how it is reflected to you - will give you grand delight.

Beloved ones - sufficient unto the time. I love you all so whole-heartedly, and it is a very large heart.

(P'taah thanks the hostess and then the host:) Dear one, our thanks, indeed. Thank you.

The host: It is my honour.

P'taah: And indeed, our thanks to you all. It is great pleasure and honour for us to be here with you. I love you. Good evening.

Chapter 6

SIXTH TRANSMISSION.
Date: 2nd of October, 1991.

P'taah: Dear Ones, good evening.

Good evening, P'taah.

P'taah: How are you all? So it is, indeed, well come to what is to be new friends. *(A number of new-comers have joined the group.)* Now, this evening we will commence to look at what some of your belief structures are. Now, what we shall do is to merely have a little chat with you, that you may put your brain in action as well as your heart, that you may bring forth questions and in this fashion you may come into clarity. So, what we shall speak of this evening is what is very often called a touchy subject, because we are going to address human sexuality. This is called a 'big number', indeed. We have spoken before in your time very much about the relationship between person to person. Now, we have touched on the relationship between parent and child and, indeed, more particularly your own childhood, because - in your experience of babehood and early childhood - you have built up your belief structures, which results in you viewing your world as the mirror to how you view yourself.

Now, humanity is sexual being, that is who you are. In your religion - we say *your* religion, speaking of that which is predominant religion of your culture - sexuality has been very misaligned, maligned - hm, maligned, that is the word. Because of your cultural heritage from childhood on, whether or not you had religious upbringing, there is within the morphogenic resonance *guilt.* We have spoken to you about 'sin'. In your culture humanity is raised to understand that what is of sexual nature is *base*, is sinful, is dirty.

It is that sexuality is not 'spiritual'. Then when you are in adulthood and you come into a relationship and then into marriage, suddenly - magically - sexuality is ok. And 'sexuality for procreation is not so very nice, but you have to do it to have babies' - is it not extraordinary?

You are sexual beings from the time of your birthing. Your body is the temple housing your light-energy. Why do you think it would be that you would be created sexual beings, if it were not considered divine expression? So, no matter how intellectually enlightened you become, dear ones, there is still within you the emotional retardation about your sexuality. Then it becomes extraordinary, because in your culture, when you reach a certain age, you begin to understand intellectually that what have been the social mores of sexuality are nonsense, that they really have nothing to do with real life. So everybody rushes out to experiment, to read whatever they can and everybody is in dreadful fear that they are not *good enough*. So you see, we come back again to that which is *not worthy*. Whether it be male or female, you look at your anatomy and believe that it is *not enough,* that it does not 'come up to standard' with whatever is socially in vogue at that present time. And then you worry about your performance. Now, nobody ever showed you how to do it, but you are supposed to know how to do it and if you don't, then you will not be acceptable and certainly not worthy of love. So, you get through those bits and pieces and then you discover through wonderful spirituality a New Age and some are fortunate enough to come into cults or sects - we would not say religions - which say it is ok to be a sexual being. Then very many will say: 'Well, it is ok, but you see, I am beyond it, I do not need it, I am involved with higher things than sexuality.'

There are many people warning about what may occur if you get too involved in anything that involves your loins. Do you not think it is a trifle extraordinary? Hm, I too think it very extraordinary. Then, you have what is a small problematical area, which is called diseasement. Sexually transmuted, terminal diseasement. Well, it is enough to put you off! Then there is marriage - forever, and 'God help you if you stray'. Dear ones, do you see the threads we are pulling in here? We are talking about expression of the heart. You know, to express joy spontaneously, the desire for communication, the knowledge that *you live in a safe universe,* the knowledge that there is no judgment - except what is in your head - the knowledge that your body is indeed an expression of your light-being, and when you create joy within your embodiment, indeed it resonates through the universes and, indeed, you live in a safe universe. And as you believe, you create, you manifest. Now, what do you think this is? It is called: *Go forth in joy and enjoy - with the heart.*

Now, we are saying how it may be. We are also saying that for eons of time humanity has used sexuality as a power-tool. The people, particularly of your culture, in these years have been using sexuality in a very materialistic fashion. Now, you know that it has been so in your history that women were sold into the slavery of marriage for money - now it is not that women are sold, but that they are selling themselves. That which is man uses his power and money and status symbols that they may be admired, and so he will take a woman who fulfils the fantasies of what is called your social consciousness, to display that he is very successful. This is *not* called: Go in joy and enjoy. And all of this is merely reflecting how you feel about Self - what you belief about Self. When you know indeed that you are a sovereign being, that you are indeed divine expression, and you go forth in freedom, and you meet somebody of wondrous resonance and you are in laughter and joy, then it is to be expected that your body will resonate that which is in your heart. And it is wondrous. If you were seeing what is occurring with this energy exchange, you would be in delight. The lights, indeed, are beautiful. When you are using sexual expression as a power-tool and when your heart is not involved, it is then to look, to see *why.* Dear ones, you are sexual beings from the moment of your birthing until the moment of your translation. It does not matter how old you are in physicality and it does not matter what the expression of your sexuality is. We are speaking of the grand judgments, that we spoke of before, of that which is homosexuality or indeed bisexuality, which is 'normal'. Heterosexuality is normal, but it is not the fullest expression. Now, we know indeed that there is much shock and horror when we say this. To give natural expression physically, where your heart is involved, is called normal, is called balance. Everything else is called judgment. - Do we have questions?

Q: (M) So how does one deal with this judgment, how does one get away from it?

P'taah: How do you think it is? What do you do with judgment?

Q: I don't know. Look at it?

P'taah: Indeed. It is first to recognize why - hm? It is to come into an understanding of what fear lies behind the judgment and then, beloved, *it is to embrace it all.*

Q: So, how do I do this practically?

P'taah: Well, do you wish us to give you a 'for instance'? - I did not think you would. Very well, when a situation occurs where you are looking at somebody else's behaviour and you are saying 'This is not right', then what you are really saying is, that you either really would like to be doing whatever but you are too afraid, or that you are merely afraid of the idea itself. So it is to look at *why*. What is the bottom-line belief underneath the judgment? You will be surprised if you may examine honestly what the fear truly is. There is great fear within man to be judged as homosexual, if he is not. We have noticed that, in this time, the judgment is not so fearful, and that those who are homosexual men are not so afraid of being found out. We have noticed that there is great fear amongst men if they are heterosexual and yet have had homosexual experience in their early life and enjoyed it, and then no longer practice homosexuality. There is a great fear that, truly, they may be 'found out'. Now, what is this? It is called fear, fear of not being normal, fear of being sinful, fear of being judged - guilt. It is not so in this category for women, because for them physical expression, not necessarily overtly sexual, is normal - culturally, do you understand? So, we would say that women 'deal' better with homosexual experiences, they do not have such guilt. *Where there is guilt, it is a lesson not learned,* and we will speak about this at another time. *Guilt is a lesson not learned.* Fear - well, it is very simple, is it not? Simply transmute the fear. And how do you do it? Take responsibility ...

Q: (M)..take off your head..

P'taah: That comes later, beloved - we must align judgment first, hm? Then you 'take off the head' and then you feel the feeling. Whatever is not love is fear. Dear ones, we say it and say it to you - *whatever is not an expression of love is an expression of fear.* How to transmute the fear?

Take responsibility for creating a situation to find the pearl of wisdom within. Align the judgment you have about yourself, about other people, about the situation. Know that indeed, all of who you are is divine expression - even the judgment - and as you embrace all into the light of who you are, so indeed you experience that which is transmutation - you feel the feeling. Is that answer enough for you, beloved?

Q: Yes, and I have another question.

P'taah: Oh, very good.

Q: The aligning of the judgment - does this mean I shall just recognize whatever my judgment is and just allow it - sort of?

P'taah: Indeed, acknowledge - allow, this brings all into harmony. This is alignment.

Q: So just the allowance of it?

P'taah: Indeed. That is how you align. Acceptance and acknowledgment of who you are, of how it is, of all, that is called vulnerability, beloved. That is called *surrender* to SELF, the surrender that you are so afraid of. That is all it is. It is most grand and it is the most powerful attribute in your universe *[surrender]*.

Q: So, if we are understanding what you are saying, then, instead of rejecting a situation, we can align everything consciously before trying not to upset ourselves...

P'taah: Indeed. But dear one, if you try to avoid the pain, of course, what occurs is that you draw it to you. *What you resist - persists.* Indeed. You do not have to call things forth as an actual experience in a form of drama and chaos, disruption and discord. But if you do not align, that is exactly what will occur, because your soul is very persistent, your soul is a little like me: *We will do anything to bring you home!*. There is no escape, beloved ones, and if it is not I, then it will be somebody else.

Q: (F) P'taah, it is to do with belief systems - do we carry fears from past lives into this life, or is it just our conditioning in this life that actually brought on the fears?

P'taah: It is a little of both. However, we will tell you this: there is indeed much hurry-scurrying into exploration of what are past lives. It is really not necessary, because whether or not the fears are from one hundred lives before or the last past lives or all of them, if you have them in this life you will have to deal with them. Do you understand? So this preoccupation with what you have brought forth into this life is really of no consequence. You could as well say that you are bringing forth fear from other probable realities of this time. And what about future lives, hm? It is all concurrent. It is all occurring at the same time outside of this space-time continuum. You can truly drive yourself 'batty', if you are becoming so engrossed in other lifetimes. We say: do not be concerned about other lifetimes, whatever the fears are - you will draw them to you. We will also say, that as you embrace, as you transmute, as you align who you are, *you are creating the change in all other lifetimes.* That is what the miracle is.

We shall create a break at this time. Dear ones, it is not so often that we see you so stunned - which is alright, you know. We are allowed to have a laugh, hm? That which is sex is not so serious, although it is the pivotal area for most of you in your life - it is the area where you have very much grief. Indeed, we will take a break. You may refresh your bodies and refresh your grand intellect and you may think of new questions, indeed.

(After the respite:)

Q: (F) P'taah, before when you mentioned sexual interaction I had the feeling that you are possibly meaning that we don't necessarily have to think of copulation when we are thinking of sexual intercourse.

P'taah: Indeed. Sexual expression - we might say in a manner of speaking - would also embrace sensuality, not necessarily that which is congress.

Q: Congress?

P'taah: Copulation. There is really more to your sexuality, indeed, than that which is copulation.

Q: So that just as much as it is okay to have sexual copulation, if we have it not, that is okay too?

P'taah: Oh, but of course.

Q: So, the people we interact with, we are actually having a sexual exchange through speaking, touching, thinking, feeling and also it is important to have a total loving relationship with ourselves, to have a sexual relationship with ourselves?

P'taah: Absolutely. You know, your sexuality is truly no different from any other aspect of who you are. So, if you are not having a full and wonderful relationship with Self, then how, indeed, can you have a sexual or any other kind of wondrous relationship?

Q: (M) P'taah, how would a more advanced civilization, for instance the Pleiadean people, handle their sexuality compared to the Earth people?

P'taah: Well, as in other aspects, it is that people - those who are humanoid, who have sexual expression - do so in the fullness as in the fullness of their normal day to day beingness. Do you understand? If the people, all of you, are in the fulness of being, in totality of beingness, then it reflects in all manners and your sexuality is only one very small part. Now, what we are really asking you to contemplate in this discussion is: what are the fears and the judgment that prevent you from being in the totality of your own sexuality? How is it that you use sexuality [as a tool] for manipulation, for power? How is it in your sexuality, that you become what you fear, in your own insecurity, in your jealousy, in your possessiveness, in your guilt, your shame - in your lack of fun? Because, indeed, as in all other aspects, sexuality may be one of spontaneity, of joy and of laughter - when it is truly from the heart. So it is this, that we ask you to contemplate.

Q: (M) Could you just shed a little light on sexual loyalty between two people who have classed themselves as a unit of specialness?

P'taah: Indeed. In marriage or relationships akin to marriage, where two people have come together - sworn faithfulness, hm? -

very often faithfulness remains out of guilt and duty. We have spoken before, beloved, of how duty becomes heavy, how it becomes resentment. So, what we are truly saying is that when you are in a relationship and it is truly from the heart, then it will be joyous. When the relationship becomes a problematical bonding and the heart is no longer truly there, then all aspects are reflected - not only sexuality. But what I am saying is: look at *how* you use your sexuality. To be using sexuality as a relief from boredom - as excitement, because your life is no longer exciting - to be used because you are insecure about how attractive you may be, how worthy you may be for the notice or the love or the desire of somebody else - do you see what I am saying? Then this is not really an expression of joy from the heart, it comes from fear. What we are saying is to look at *why*, always to look at *why*. And do you know, truly, there is no limit on how many people you may love. You see, there are so many issues which are bottom line, when we are speaking about sexuality. When there is truly a bonding between two people, it has nothing to do with duty. When people are truly aligned with SELF, then there is no jealousy and no person may own somebody else. When you are possessive, it is because you are terrified of losing, do you understand? We are not for a moment suggesting that humanity should drop everything and race off to experience as much sexuality with whomsoever and howsoever, hm? We are asking to look at how it is with you in the manner of sexuality, that does not bring you joy, where you do not feel fulfilled - because you are not in understanding of who you are. Does this somewhat answer your question, beloved?

Q: Yes, thank you.

P'taah: We are saying that when you are making vows to love somebody unto death, it is slightly unrealistic, when you cannot even love yourself unto life. In fact, what you are all doing is *unloving yourself to death*. So it is, to take sexuality and bring it into the totality of your being. We are speaking of balance. We are speaking of *masculine - feminine energy in balance*, within and without. We are asking you to look at the games you play, and that is alright - indeed, it is alright. It is valid, it exists, but it is not necessarily the fashion which will bring you the most joy. - Questions?

Q: (F) When vows are exchanged, if there is a feeling that something is limiting you, therefore it is not desirable, should that be a sign to you that something is not alright?

P'taah: Indeed, but usually, beloved, it is not to look outside to the relationship, but to look within to see what is limitation. Because, you know, *there is only one thing that is limiting you and that is you.* When you feel constricted in your relationship with another, it is to look at the constrictions within you, because it is co-creation. Then indeed, when you are perfectly satisfied that you are very clear about how it is and if it is not satisfactory, then you may say: 'Beloved, it is time that we are not together' or 'It is time that you are to look at how it is for you'. Do you understand? To discuss, to have communication between you that you may be very clear, that you may take responsibility. To know that *you* have created the situation, and if you have created somebody in your life that is causing limitation, then *know*, dear one, that it is *you* who has drawn into your life a mirror of how limited you are. It is also to know, when you are clear and when you are aligned, you will not draw in anybody who is reflecting that limitation. Is it clear?

Q: Yes.

Q: (F) I do not have a partner at the moment and I feel like my work or myself is balancing my male and female and for most of my life my female side has been predominant and I have not been assertive enough. In my last relationship I was learning to be more assertive with my feelings. My question is: I am content within myself and may be this is one reason why I do not have a partner at the moment, or am I just preparing myself for the right one to come along?

P'taah: Dear one, the *right one* who will come along is you. We will say that at this time there are many people who are not in a relationship. That is very natural, because you are all so busy finding out who you are, that you really do not have time - and that is alright. Now, it works on many levels. There are of course those people who are afraid of being in a relationship, and those of adult years who have had one or many relationships, decide that exploration of Self - and that which is New Age - is a very good excuse not to be involved, *not to be hurt again.* There are those who are busy scurrying about looking for a partner and cannot find one; and we will remind you that the harder you look, the less you will find, because - indeed - it is a reflection of how it is within you. It is to allow.

Now, that which is feminine energy is called allowance. It is also for you to understand, that as you simply *know,* that everything you require will come forth to you, *without doing anything,* you will create wondrousness in your life. Simply by the *knowing* that *all, all* that is required for your growth, for your joy, for your contentment, *and indeed, for those lessons not truly desired by you,* will come forth in the ripeness of time.

Q: (M) Are there any other beings or persons in the universe who have sexuality as we have here on this planet?

P'taah: Indeed. Many, many humanoid beings, many. Now, there are some people who are humanoid, who do not use sexuality, as you understand it, for procreation. There are some who express sexuality in a slightly different fashion. We are not saying that they do not have all the moving parts, but just simply, that they have a grander fashion of energy exchange. And, dear ones, do not forget, that what sexuality is between two people - no matter what sex, - is *energy exchange.* And as I have said before to you: imagine, what it would be like if you were able to see the lights of energy that you are creating; not only sexually, but when you are in grand anger. What we are saying is, when there is great emotion, indescribable fireworks display.

Q: In the aura?

P'taah: Indeed, but when you think of auras, dear one, you think of soft lights about you. We are speaking of grand sparks - rockets indeed. There are many races of people, humanoid, in other parts of the multiverses, and indeed there is another species of humanoids living within your planet, that are not normally seen. We are speaking of a grand race of people who inhabit what is called your inner Earth. It is also a different space/time continuum, a different dimension of reality.

(A gentleman enquires about the sexuality of the inner Earth people:)

Q: The sexual behaviour of the inner Earth people, their sexual understanding, is it similar to ours, although they are on the etheric level?

P'taah: Indeed. Hm, we are not saying they are on the etheric level, beloved.

Q: So, please, do say where they are or what they are.

P'taah: We are saying that they are the people of the inner Earth that are simply not of this dimension of your reality. However, their dimension of their reality is as non-etheric as yours. But they are certainly more in knowing of their own etheric. It is a grand, wondrous civilization. Older than, hm, much, much older than your own history of humanity. We are not speaking of the pre-history of your planet. We are speaking of your history of time. Grand civilization, very advanced in technology and always having been in communication - physically and mentally - with the star people. In a fashion like humanities of other worlds they are travelling from galaxy to galaxy. They are travelling *within dimensions*. You see, very often, when you are thinking of dimensions of reality, you are thinking in the vertical. You are thinking of yourselves as you poor, lowly humanity of third density, bottom of the heap. Is it not so?

(Amused, the audience agrees.)

Indeed, this is called *not worthy*. These other harmonic frequencies have a valid reality, physical realty, as do you. There is a different vibration, and you know when we say 'higher vibration', we are speaking scientifically. Not that they are higher echelon[1]. There are certainly many, not all, more knowing in the technologies, and they are, of course, more knowing in technologies because they are more knowing of how the humanities, how the universes, how the light works. It does not make them grander than you, because, truly,

---

[1] French, from the Latin word scala, a ladder, implying a hierarchical structure.

beloved ones, there is no separation. All is divine expression - it is just *different*. Do you say which is grander, orchid or daisy or rose? Indeed, you may have preferences, but when you live in this part of the world, where grand orchids grow like weeds, very often you are preferring the rose. And when you live in the continent of Europe, where an orchid is very exotic, you would perhaps say it is far more special than the rose, which grows [there] like a weed. But all, all is divine expression. In the energy of the Source, of the All That Is, there is no judgment, there simply IS.

Q: (M) P'taah, is there a way to visit these different space/time continuums?

P'taah: Dear one, there truly are many ways and, in a fashion, very often you do. It is that you do not have a conscious memory. Now, we would say, [to visit them] in your conscious fashion, the time is not yet ripe, but the time is coming very soon. Dear one, that is what we have been speaking with you about. As the consciousness of humanity comes into expansion, as you come to know more and more of who you are, as you come more into allowance, acknowledgement, embracement of who you are, so you are indeed changing the resonance of the whole of humanity of your planet of this time - so you are in alignment with the other changes which are occurring in the multiverses, galactic changes, cyclic changes that we have spoken to you of. When these changes occur, you will travel in your consciousness, physically *and* where you do not need to take your body, but certainly in consciousness you will be travelling in time, outside of what you understand as time and space. Travelling through universes, through galaxies.

Q: I feel it is possible, what you just said. I feel it - within. Yet, it seems so far away.

P'taah: Indeed. You know, in a way it is far away, because you have not consciously experienced it - yet. And so it is like a child, who knows he is going for the first time to a circus, and he is counting the days and he knows it will be wondrous and exciting, and he has not seen the movies, but has heard the stories. He cannot truly imagine what it will *feel* like, but in excitement and expectancy it seems so long. But you know, beloved, in another way, you are saying how the time is going so fast, that you are in warp speed, all of you. And you say to each other with a shake of the head: 'Is it not extraordinary - the days go so quickly, there is no time any more.' So, in a fashion it is the yearning for the circus and wondering if it will ever come - or, indeed, if some calamity will occur, that your vehicle will break down, or that your father will be angry and you will never get there, hm?

Q: Is this yearning bringing it about faster?

P'taah: Dear one, everything has its correctness of time. You cannot hasten that, which is already 'rigged', so to speak. It has its own timing. We may say that everything is on line.

Q: So, just allow?!

P'taah: Indeed.

(A lady seeks clarification on the subject of sexual diseases:)

Q: P'taah, back to sexuality.

P'taah: Indeed, beloved.

Q: Just for further clarification: When we are in joy and we act from the heart, then there is no fear of sexual diseasements that one can 'catch'?

P'taah: We speak specifically about AIDS, because that is one of the great fears of your time. Now, you know it is all multi-level. Certainly, if you go forth with joy in your heart, it is wondrous resonance, but it is always to understand bottom line belief, core-belief. If there is no guilt, if there is not a sense of unworthiness and if it is truly joy from the heart - dear one, how on Earth can you produce dis-easement? Now, we have said this forth and we would like to bring forth another point and that is: The children on your planet, who have contracted this diseasement by transference of blood, either in the foetal stage or in early childhood - and you may indeed say: 'Well, the child has no guilt etcetera, etcetera..' - dear ones, we would remind you, that you come forth on the planet, lifetime after lifetime *for the experience*. Nothing is written in stone. You may change everything according to your beliefs. It is very important that you understand this.

(P'taah gazes for a little while at a lit globe of the Earth in the room and remarks:) It is not a true representation, you know? It is the wrong shape. Your Earth is not round like an orange.

(Another lady expresses her appreciation:)

Q: P'taah, I would like to say that in these sessions I really have been feeling such a unity amongst us all, and it seems to me very strongly that you are acting like a pyramid energy helping us to focus on the issues we have to learn and that you are helping us very, very much. I feel that very strongly, thank you.

P'taah: Dear one, it is a joy and as you envision it, so it is really, in a way, that you are using this energy, if you like, as a conduit. And if you wish, then indeed you may bring forth all of the queries for your own clarity - that is why we are here. As you do so in this course amongst each other or, indeed, with this energy, you are creating something quite wonderful.

Q: During a little expedition to the country the other day I really felt I was not quite sure how you see this world, if you only see through the eyes of the vehicle. I just felt very much I wanted to share the beauty of the country with you, and I have been doing this on a few occasions. Can you - can I share something with you by doing that?

P'taah: But indeed. We have said this before, dear one, that as you call forth the energy, so it is. And you well know, indeed.

Q: (F) P'taah, concerning sexuality and sensuality: How can we educate our children into freedom of feeling, while we are still struggling with the limitations?

P'taah: Indeed, beloved, perhaps it is the children who will be teaching the parents. It is called allowance, that is all, to love and to allow. To allow the children natural expression and then they may teach you many things - not only about sexuality. And children are very sexual little beings. They will teach many wonderful things, more and more in these next years, because truly, *many, many children born at this time are shining lights, who come forth to help the transition of humanity.*

Q: (F) P'taah: With the transition that will happen, will the sexuality still be the same or will that advance with the shift?

P'taah: It will change - all things will change, all expressions will change.

Q: Right - and is this the fourth dimension?

P'taah: Dear one, it does not matter what it is called. We will say fourth dimension. Some people say fifth dimension, but as you know no other dimension than this one, it does not matter what name it is given. Let us not get hung up on the name of it.

Q: (M) P'taah, regarding belief structures - it seems to me that certain beliefs are very ingrained in our human society. Now, if we truly would like to change certain beliefs - how do we do it?

P'taah: Dear one, first of all you have to identify them, because the beliefs that create for you the most problems are the ones you really do not understand that you have. You see, it is very simple to identify beliefs, when you say: 'I believe that the government are all fools' or 'I believe the sky is blue.' But we are speaking of situations, where you emotionally react. Then it is to look to say: what is beneath this reaction, what is the bottom line fear? It is in this way that you discover beliefs, which you did not understand that you are holding. Even beliefs of which you intellectually know, that do not help you. It is like baggage you are carrying around, beliefs that are really of parents or society, which have nothing to do with you. Do you understand? Once the belief is identified it is simple, because you can look and say: 'This belief I am carrying around no longer serves me'. By *embracing* that belief you create the change, because very often the intellectualization of it does not change it, but it is the first step. When you recognize [the belief], then you can say: 'Well, it is crazy', hm? Then you will come into the knowing, the acceptance, the embracement; to say: 'It is alright to believe this, but now it is time for a change', and then it is to very gently allow the change to occur. You will be very surprised at many of the beliefs that you think have to do with your social structure, and they do not at all. They have to do with how unworthy you are. It is very simple for a [belief] structure, where it fits in with your bottom line fear, to hook on to it and there it stays - then you have in your conscious mind justification, do you understand?

Q: I think I do, however, I need to recapitulate later with the help of the transcript, I guess.

P'taah: Very well. Would anybody require that we speak more of this? Are you all in understanding? Very well. If a situation occurs in your life and you have an emotional reaction to the situation, we will say anger - we speak of sexuality and you have just come back to find your beloved in the arms of another and you become very angry and throw a pie-dish and you storm out of the house, drop into your vehicle and go to your mother. Now, your mother will say: 'What a bastard.' Then you say: 'Yes, but I did a terrible thing - I threw the pie-dish, I shouted.' And your mother will say: 'But dearest one, it is perfectly justifiable. Had it been me, I would have hit him on the head with the pie-dish.' So, you are comforted and then all your friends say: 'Oh, dear me, what a swine.' But, you see, dear one, what we are saying is that the bottom line is called desolation, betrayal - which is fear - the fear of not being enough, that your beloved will turn to another. It is called jealousy, the fear of not being enough. It is called possessiveness, the fear of not ever possessing. You understand? So, when it is that the reaction will fit into a social structure, very often you do not feel that you have to go to the beliefs. If you are to transmute the fear immediately, you would never have to deal with [such] situations again. And dear ones, it is to remember always, that *everything which is not love is an expression of fear*, which may be transmuted. In that same sense it is not necessary to intellectually know what it is. It is to say: 'This is how I feel - or do *not* feel, really - I am in pain, I take responsibility, align the judgment, put the head under the arm and feel the feeling'. But you see, you are intellectual creatures. You want to have order in your universe. You want to know in your mind how it is all created. You know truly, all of these words we are saying can be condensed into very little and we are not telling you anything that you do not know, because in truth you know it all. All we need to say is that everything about you is divine expression and when you are in pain, feel the feeling. *(P'taah turns toward the transcriber of the material and remarks:)* But that would not write the manuscript!

(A gentleman, who attends these sessions infrequently:)

Q: Could you just explain a little more in practical terms about feeling the feeling? I am not quite clear on that.

P'taah: *(Teasing)* That is because you do not come to see us all the time, beloved. Do you know, it does not matter how often we say it all, and dear one, you have only brought it forth because there are many who need to hear it again. Humanity knows not very much about feeling and pain is *not* feeling, it is *resistance to feeling.* When you are in joy - that is feeling. This is a very brief recapitulation. To align the judgment is merely to acknowledge and accept, to surrender into who you are and *that automatically creates alignment - balance,* do you understand that? — Not very well! — Pain is created by judgment, *but you judge the judgment,* let alone everything else. While you are in judgment, there will be pain. We are speaking of specific situations in your life. To release the resistance of the pain, it is necessary to, if you like, nullify the judgment.

Q: Do you mean: To accept, yes, I judge that person to be wrong and it is okay to judge them to be wrong. So where do I ever get off the judging of people to be wrong?

P'taah: What we are saying, beloved, is that when you make judgment wrong, you will never be rid of it, because whatever it is that you are resisting, whatever it is that you invalidate, whatever it is that you try to push from you - *the universe, without judgment, will bring forth that which you are empowering.* The only way that you create the change is to say: 'I desire the change, I understand that the judgment of myself, of the situation, of the people involved, is valid. It is part of my being human and my being human is divine expression. But now I desire the change'. And it is to gather the judgment into Self, with love, with compassion. In that fashion the judgment is aligned, which creates the release of the claws of resistance of pain, so that the energy chakras are open, so that it [feeling] may pass through into the heart to become ecstasy. Is it somewhat clearer? Whatever it is in your life that you wish to align, it only means to bring it into balance. To become neutral, if you like. You see, you may have great pain, you may express the pain in your rage or with your tears, with anger. You may suppress it all, not give forth any expression of your pain. The balance is that you *align* the judgment, do you understand? So you create the change.

Q: So, basically, just accept everybody for what they journey, without judging that - yes, it is clearer.

P'taah: Other peoples' journeys are none of your business. What anybody else thinks of you is none of your business. Your business is you - Self. As you can come into love and compassion, acceptance and acknowledgement of your own divinity, how can you not be thusly with all people? — We will have one more question?

Q: (F) With regard to what you have just been saying, I have been working with this alignment. I just wanted to get it clear with you whether it is the same as what you are saying. I use this position (the lady indicates a meditative position) and go into the feeling of whatever emotion has caused some discord at the time, just focusing on the situation and going into the feeling and then changing the picture in my mind into a positive one of how I want to feel.

P'taah: Dear one, that is very valid, indeed, but it is not transmutation, because transmutation is not *done*. You cannot *do* it - you can only *allow* it to occur!

Q: So, you are saying that it is a waste of time?

P'taah: Dear one, we are not saying that it is a waste of time. It is wondrous that you may put forth pictures to be positive in your mind. It will bring you great relief, certainly. What we are saying, really, is that *the only thing that will create the change is embracement.* We will say again, *whatever you invalidate you empower.* You see, we say again: the universe does not judge. There is no good or bad. There is no right or wrong, there simply IS. So when you want to create the change, by all means create a picture in your mind of how you wish it to be. It is always, of course, thought creating reality. So it is wondrous to do. But we are speaking of bottom line and transmutation is bottom line, because that is the true miracle. Now, you have not been here for discussion of transmutation. We will not [discuss] at this time, because it is time for ending the session, really, and we will not be speaking again in the, what you would call lead-up, about transmutation, because these discussions are very ordered for the manuscript. So we are constrained. However, you will be able to hear about it and to read about it. We only wish that you will *know* about it. But do not worry, nobody here is really knowing about it. *(Laughter).*

But that will change, and you will all know about it. To *know*, in the breast of each and everyone of you. To create the flower and within the lotus the jewel. And the jewel is truly the knowing of your own divinity, the knowing of the GOD I AM. That is how it will be for you. *(P'taah thanks the host now:)*

Dear one, thank you.

Q: Thank you, P'taah.

P'taah: Hm, indeed. *(And to the hostess:)* Our thanks, beloved. *(And to all:)* Our thanks, indeed to you all. Has it been quite riveting?

(The audience agrees heartily.)

We thought it might be. *(P'taah says with great tenderness:)* I do love you all dearly.

(A lady:)

Q: We love you too.

P'taah: *(Almost whispering:)* But I know, beloved woman.

And that what you are makes a very beautiful picture. I wish indeed, that you could all see who you really are at this moment and know how beautiful you are. Our love and thanks, dear ones. Good evening.

Chapter 7

SEVENTH TRANSMISSION.
Date: 9th of October, 1991.

P'taah: Good evening, dear ones.

Audience: Good evening, P'taah.

P'taah: Well indeed. A pleasure to see you all, all. Grand journeying indeed, dear ones. You are well come to this place, indeed. *(Turning toward a newcomer)* And so, dear one, how are you?

Q: (M) Very well, thank you.

P'taah: We are very happy that you have come to this wondrous place and this energy of healing, of great gentleness, so you may find indeed within you that it is to be gentle with who you are and to heal the pain. Gentle. Nothing to *do*, dear one. We are pleased that you are being gentle with your body.

(P'taah moves through the audience and addresses the daughter of the hostess, a young lady, who had just returned from a three year sailing trip around the world.) Hello indeed. The seafarer. That which is the sea has many grand lessons. It is from the sea that it is so easy to learn how to be in harmony with awesome nature, indeed?

In the last week, we spoke to you about the sexuality of the humanity, and indeed we understand that in these days since, many of you have taken pause to think about your belief structures regarding your sexuality. You know, guilt has been programmed into you, not only for eons of your time in this period of your history, but indeed instilled in this lifetime. In this age of great allowance, of great freedom, this time of independence of the humanity, of this culture, still you carry guilt. Now, much of your guilt has nothing to do with sexuality, but it seems that as we have talked about your sexuality, it is a good time to talk about your guilt.

Now you know, guilt, in truth, is merely a lesson which has not been learned. Regret is also a lesson not learned. We would ask you

115

to consider this: when you are doing what is called a grand review of your life, if you would take a step back, to have overview of your life thus far, then you bring forth the intellectual knowledge that at soul level you bring forth lessons every day of your time; to know that, as every facet of who you are is indeed, a reflection of divinity, divine expression, that there cannot be a wrong decision, that there is no right or wrong, there simply IS. How then indeed you be in guilt? How indeed can you have regret? It is very simple, is it not? *Each situation that arises in your life, every situation that you judge to be discordant, that brings you unhappiness, is there for you to learn.* [There is] no right and no wrong, beloved ones. There simply IS. *Know, that you are GOD I AM.* Dear ones, you are not lowly humanities, third density beings struggling to be spiritual. You are all spiritual beings, else you would not be here. You are all, *all*, grand spiritual beings choosing to experience this dimension of reality. It is a courageous choice. One of the reasons you choose to experience this dimension of reality is indeed the intensity of the experience, the colour, the vibrancy, and you have chosen every experience in your life, every one. Now, from the state of your present great awareness and great knowing, and great understanding, you may look back on your years, and you may say: 'My God, however did I do this? How could I have been so stupid? How could I have been so awful, wicked, uncaring, etc. etc. etc.' We could be with you all of this evening and list all of those attributes that you castigate yourself with. Beloved ones, *the reason is because you wanted to*, because you chose to for the lesson to be learned. It is so simple. So in truth, why would you feel guilty? Guilty of hurting somebody? Indeed. But that somebody was in there with you, co-creating every situation. Indeed, dear one? *(P'taah looks upon a gentleman.)* How are you going with your guilt trip?

Q: (Great laughter) I'm doing fine.

P'taah: Are you indeed?

Q: Plenty of judgment.

P'taah: Hm, very good. *(Then, mocking outrage:)* Not judgment, dear one. Such spiritual being as Self, in judgment? Good grief.

116

Yes, beloved ones, everybody was in there with you fifty percent, for their own learning. And who are you to judge how somebody else would create their own learning, whether it be your child, your sweetheart, parents, or merely that which is casual acquaintance. Now, also we have been with people who have felt they have caused the death of somebody on occasion; in an accident of your vehicle, or it has been a demise caused by a failure of the physical body. *There is no such thing as accident.* When the one driving one of your very sophisticated vehicles would have an accident and cause the death of a creature, animal or human, it is for you to know - indeed - that it was *chosen.* So where is the guilt? And do you know, regret is truly as ridiculous as guilt? How can you regret? You would not be where you are at this moment, if it were not for every situation you have created within your life. And we understand that most of you are not necessarily all that happy about where you are at this moment. However, would you truly have it any other way? I will tell you that you would not, else it would be another way. The next time that you catch yourself with a very severe attack of guilt, or a severe attack of regret, we would ask you to recall these words: *you have chosen and there is no such thing as wrong. There simply IS.* Now, do we have a question at this time? Indeed, dear one.

Q: (M) P'taah, please explain the mechanics of this fifty-fifty percent mechanism of choosing a situation like an accident, for instance.

P'taah: We understand your desire to understand the mechanics of co-creation. Is it truly so difficult to understand? If you are co-creating something joyous, it is not difficult to understand, hm? You see, beloved, it is only your judgment that makes a difference in what you judge to be dire circumstance, or wondrous circumstance. All is for the lesson to be learned.

Now, we will talk to you again about your power. Your physical reality, indeed, the reality of this density is created from your belief structures, hm? And in your mind you put forth [thought] according to your beliefs. The thought goes forth, embraced by emotion, into the universe. And beloved, the universe has no judgment, whether you are in desire of something that you judge to be wondrous, or something which you judge to be discordant, or bad - the universe does not judge. It simply re-arranges itself to accommodate your belief. So, a person who drives a vehicle may believe that he is a person who would inflict injury upon another. We do not mean consciously, hm? And, as you know, some of the most powerful beliefs are not of the conscious mind. Or a desire to be punished for something in the past, inflicting injury and what-not; a desire to be punished. Now we are speaking hypothetically, beloved. And there may be another person who desires to translate, for one reason or another. It may be that the world is too difficult. What are the myriad reasons that people desire to translate from this dimension to another, hm? And so, given the probabilities, the two will meet, and it is called accident. And in this fashion, each creates what is desired. Do you understand how it works? Now it works thusly, when people meet and fall in love unexpectedly. And they say - is it not extraordinary, I was not looking for this. And around the corner, in the supermarket, there - the two people's eyes will meet and it is wondrous love. Is that not how it may occur? So it is that the people have co-created for whatever reason, and will continue to do so for as long as they are together, to reflect that which they judge to be wondrous, and that which they judge to be discordant to each other, that each may learn. Does that answer your question, dear one?

Q: Yes, it does. It does, however it raises another question.

P'taah: But of course.

Q: You just said that the emotional embrace of thought is the fundamental force..

P'taah: It is the empowerer.

Q: Okay. Now I can imagine myself embracing, with great emotion, a lotto win?

P'taah: Indeed.

Q: Yet, it seems to be very illusive, so why does it not occur?

P'taah: Because: the bottom line is, you do not believe that you deserve it.

Q: I don't think so.

P'taah: Do you not, dear one? Because there are other methods, more beneficial, for you to understand about abundance.

Q: Yes, that makes sense.

P'taah: Indeed. It is really most simple. You know, very often, when we are speaking about this creating your own reality, people will say: 'But where is my abundance?' We spoke before about this. What do you really believe about money? Now, we would say this to you: Whatever situation in your life you believe not to be harmonious - take your paper and make a list. On one side of your paper write down all the wondrous things that you feel about the situation. On the other side write down all the things that you think are not so good. That is positive - negative. You will be very surprised. You may also use this method to discover what it is you truly believe about yourself. And in this fashion you may write down all of the things that you really despise heartily in your heart or dislike about other people. When you have finished your list - and you may be very personal, you may write down all the things you heartily dislike about your mother, or lover, or the person who lives next door - when you have finished your list, examine it carefully, because all of the things you really dislike about others are all the things you heartily despise about Self. Because, beloved ones, we will say again: There is nothing outside of who you are, that is not but a mirror, a reflection. And so it is that you may truly come into an understanding of what grand, powerful masters you all are.

Q: (F) If I see somebody out there who is doing something I do not like to see, it does not mean necessarily that I am doing it now, but if I would do it, I would not accept it in myself?

P'taah: Indeed. Now, we would suggest this to you also: If you see somebody outside of Self, who you judge to be very cruel - somebody who is beating an animal - would that not rouse great ire within your breast? Indeed it would! And you would say: This is not a reflection of me, because I have never beaten an animal and indeed, I love them very much and I would not do so'. However, *we would ask you to examine the cruelty that you practice toward yourself.* Do you understand?

Q: Yes.

P'taah: Questions?

Q: (F) P'taah: I don't think I really understand the answer to that last question. How can you relate the beating to yourself, when you judge someone who is cruel to animals.

P'taah: Dear one, really it is the *cruelty*. Whether it be somebody who beats a child or the grandmother or the dog, what they are displaying is *fear*; and in that surge of fear there is a display of what you judge to be cruelty. Now, we are remembering - are we not - about co-creation for the [learning of] lessons. So it is really to look at what is judged to be cruelty, and if you cannot identify any occasion where you have been cruel in this lifetime to another - and we would be most surprised - then it is certainly to look at that which is cruelty to yourself, because you are [cruel to yourself], all of you. Do you understand? And it is also to know, dear one, that it is all on so many levels. What you are engaged in at this time can be said to be 'peeling the skins off the onion', revealing to yourself more and more and more of who you are. To embrace who you are, to acknowledge that this indeed is the GOD I AM in every facet. *In the embrace, in the acknowledgment* you create the change that you so fervently desire.

You will not create change by trying to hide from who you are. You will not create the change by invalidating those facets of who you are, whether you invalidate those facets that you see reflected in another, or whether, indeed, it is truly recognized as yourself. It is to know that *whatever you invalidate you empower.* That is called *universal I am.* As you invalidate your fear, invalidate your pain, so you bring it forth. The universe does not judge, dear ones. *Whatever it is that you are shooting from your being into the universe comes back to you.* We remind you that thought does not stay locked in your head. In fact, in a way it does not even reside in your head. That is why you do not miss it, when you take it [the head] off and tuck it under your arm. *(Amused response.)*

Q: (M) How then do you validate a fear?

P'taah: You acknowledge that it is.

Q: You acknowledge that you are fearful of that circumstance?
P'taah: Indeed, and it is to accept that the part that is fearful is valid and a divine aspect of who you are. In this way then you may embrace it into *who* you are. In that embracement the change is created. In this fashion, beloved ones, if you do this and recognize the fear as it occurs, you need not experience it [the fear] in physical reality. If you invalidate the fear, if you push it from you, suppress it, hide from it, then you are putting so much energy into this, that that is *exactly* what you will attract.

Q: (M) P'taah, I understand then, that probably a good way out of this demise would be to step back and to watch oneself in a manner detached from the situation, like an observer? Is this the way....?
P'taah: Indeed. It is a way. We have spoken of this in terms of your anger. We have said: do not invalidate the anger, and 'if you do not catch it - go for it'. However, when you have experienced the full joy of your fireworks and display of anger, it is to say: 'Wherein lies the fear?'. It is to identify [the fear]. Then it is not to say: 'My God, this is so terrible, we must put it down, suppress it', etc. It is to say: 'Indeed, that is who I am - I am terrified'. Acknowledge, have a little compassion. Accept that every facet of who you are is valid and divine, else it would not be. Then it is likened unto a surrender to it, an embrace, and in that you are creating the change, the flowering, the unfolding.

Q: This is wonderful.
P'taah: Indeed it is wonderful. But dear ones, so are you.

Q: (F) P'taah, in regard to co-creation: Do we create as we go along through this life, or do we do it before we come actually on to this planet?
P'taah: Well, it is indeed in several parts, beloved. Because, before your birthing, you have chosen what we would call the 'broad spectrum game plan'. That is, that which be parent, that which be sibling, and often you will create in your life a coming together with others who are very dear to you from many lifetimes. You choose your cultural environment etc., you understand? There may be an

overall choice of what is to come forth in the lifetime and lessons to be learned and joys to be experienced in light of what you are experiencing in other lifetimes. It would be dreadfully dreary if you would be doing exactly the same in every lifetime, because that way you would not have the variation, hm? So, you lay down the game plan and then you come forth into the physical reality of the lifetime and moment by moment you create how you may play it out. Dear ones, we say this to you also: as you come into your knowing, and as you unfold into transmutation of the pain and anguish of your life, so you are *affecting every lifetime*, not merely this one. And so you are also re-writing your own history of this lifetime; you are changing what is called past.

Q: (M) Is it possible to get in touch with the purpose of a particular lifetime? Would it be of any value anyway?

P'taah: Well, dear one, the purpose of your lifetimes, of this lifetime should be - by now - fairly clear to all of you. It [the purpose] is common, as you have created yourself into this time and space of your planet's history and the history of the humanity of this planet. So perhaps we would say that *the purpose of this lifetime is to become God realized.*

Q: But each person is confronted with special lessons?!

P'taah: Indeed, and you present yourself with them day by day.

Q: Is it possible for me to look ahead to see what I have to learn to help direct myself? No - I answered it already!!

P'taah: Indeed, and on that note, beloved ones, we will take a break, so that all of you ascended masters may refresh your bodies. - It is very nice to see the smiling faces, you know? It is called: lighten up and have a ball. Very well, dear ones, we would ask for silence for two minutes.

(After the break.)

P'taah: And so dear ones, we will continue. Questions?

Q: (F) P'taah, how can we connect with those parts in ourselves that we have cut off in the past because of our judgments?

P'taah: Dear one, do not worry too much, because as you become more and more within the quickening of what is occurring - not only within the consciousness of humanity, but within the Earth itself - so you will draw to you situations that will bring all the judgment and all of the fear to the fore. And it is then, moment by moment, that you may regard to align everything in your life that is causing you to be in pain and in fear. Dear ones, it is not to do anything, it is simply to *be* and to allow the soul energy and your belief structures to bring forth day by day, that which is for learning, for the aligning, to create the harmony. It is not that you may rush about to do things, to make it better, hm? And it is also, beloved ones, for you to remember that as you are desiring to be *better*, you are invalidating what you are now. What you are now indeed is perfection, in each moment is perfection. That which is perfection, as we have stated before, is not what is called some finished product that goes with your Ascension, hm? What you are at this moment in your own divinity *is* perfection. How else could it be, because who you are, indeed, is an expression of divinity in all its myriad of facets.

Q: I hear what you are saying. I'm contemplating it.

P'taah: You are in judgment, eh? *(Laughter.)*

Q: Does that mean that in actual fact there is nothing to strive for?

P'taah: Oh, indeed it does.

Q: So it is actually a misapprehension, I mean, we have this strong programming that we're not perfect, even though you say we are; we have such a strong programming that we must be better?

P'taah: Indeed. Do nothing!

Q: Just live, just live - don't even think about being better? Is that alright? Because I worry... So we should not worry, thinking too hard about how we could be better.

P'taah: Exactly, beloved woman.

Q: It is not easy.

(Sympathetic laughter.)

123

You see, beloved, it is all so simple: *there is no thing to do.* What you are at this moment is called divine expression. It is to BE - human BEING. *Being* in every now moment, *knowing,* that if you are living in the fullness of the moment - without striving, without judging that it could be better - that you are fulfilling divine essence upon your planet. If you are truly in the now, then you are not in judgment, because you are not comparing with yesterday, and you are not worrying about tomorrow. Everything that you have ever been and everything that you ever will be is of the All That Is. You are so busy worrying about the changes to come, to fit all the pieces of the puzzle together, trying to change what has been. Worrying about your abundance, worrying about your relationships or lack of - and truly: *just stop doing and start being and know, what you are is the thought of the All That Is.* And there is no part of you, beloved ones, that is judged by anybody. There truly is no grand role model for you to compare yourself with - *nothing* to be afraid of, *nothing* to be ashamed of. *(Very softly:)* Know that always you did the best you knew how. *There is nothing to do.* Each of you is a jewel, each of you is to be regarded with the awe, as you would regard the stars in your firmament, each different and each breathtakingly beautiful. Nothing to *do.* We are certainly in understanding of your desire for change and that is wonderful, but the change occurs with the embracement, with the acceptance. So you come, little by little, into the true knowing within your breast that you are the thought of the God/ Goddes, of the All That Is. It is not to struggle to be spiritual - you *are* spiritual. It is not to struggle to be better, you are all perfection - truly *awesome* in your grandeur.

Q: (F) P'taah, are you saying that we should allow life to happen more, rather than to be planning it too deeply?

P'taah: You know, as you lay down your very rigid plans of how it should be and as you see in your minds eye the timing of how it should be, what do you think is occurring?

Q: We are holding it in place.

P'taah: Indeed, what you are doing is to shut down, as it were, all the myriad possibilities of what may be. The probable realities, hm?

The combinations. What you are doing is limiting your own creativity. Now, it is simply to say: 'This is what I desire from the God/Goddess of my being - this is what I desire to create'. To put it forth into the universe, to forget in the *knowing* that what you desire *will* occur. The moment you decide *how* it may occur and *when* it may occur, then you are shutting down your own creativity, you understand? So, certainly it is [alright] to plan and say: 'I would desire to do such and such', but it is not to be rigid. It is to allow the flow of creativity, that everything may flow from your own divinity. Indeed.

Q: (M) P'taah, I have a few questions here. I find my mind running around in circles at times, because I am not sure of the basic mechanics of what we are talking about. Could you elaborate on some of those points...?

P'taah: Indeed, dear one, let us first say that the 'basic mechanics' is simply that *you are God*, that you are an expression of divinity - a thought of the All That Is, if you like. As we have said before: you are God smelling the rose. What I mean by this is that you are the All That Is *expressing* in this dimension of reality in this wondrousness. That is called the 'basic mechanics' of how it is. And when you *know* this, indeed, you will know all else. But do continue, dear one.

Q: Could you elaborate on what space and energy is, in terms which we can understand?

P'taah: No. You see, dear one, when we speak to you about energy you truly do not understand, because *all* I have been speaking of *is* energy. And you do understand that everything *is* energy.

So, we say to you that the source of energy, in whatever form it manifests, is simply the Source, the All That Is, the unknowable, dear one. We cannot give you a definition of God, except by the definition, that all you need to do is look into the mirror. As for space, in your terms: we could say that space would be a mathematical equation, if you are being scientific. But you see, you do not truly want to know what space is, you want to know *how* it occurs.

Q: Right.

P'taah: I do not believe that any definition I would give you would be more comprehensible than anything that you have read. You are asking 'How did the galaxies occur?' Indeed?

Q: And where everything came from.

P'taah: Indeed, but we have told you. Dear one, in truth, that what is mathematical equation is only a map of what is already there. There is no mathematical equation known to humanity which describes infinity in the pure thought of it, because, you see, the pure thought of it is the Source, the God/Goddess of the All That Is.

Q: Because it is beyond us to comprehend...

P'taah: Indeed. Now, in a fashion you could say that the impersonal God Being is unknowable to all, except through reflection. And I am speaking about beings who are - in terms of dimensions of reality - far, far beyond this dimension. You could in fact say, that you may know the divine Source through that which is emotion. That which is love, dear one, that which is lack of separation. *That* is the knowing of God.

Q: Could you, as well, elaborate a little on how matter is actually formed from energy?

P'taah: Well, you see, it is consciousness that creates matter. That is the missing link of your quantum physics.

Q: So matter is instantly there and instantly not there?

P'taah: Indeed. Matter is only a frequency.

Q: How does that relate to thought then, if thought is energy as well?

P'taah: But indeed it is. You may say that thought is of higher frequency than light.

Q: And everything below the frequency of light is consciousness?

P'taah: Everything *above* the frequency of light. You see, it is a misapprehension that your scientists have lived with for quite some time, that nothing is faster than the speed of light. It is not so. And when your scientists come into the knowing of travel throughout the galaxies, they will understand that to travel in hyperspace[1], one

---

[1]   Hyperspace, an imaginary space beyond the three-dimensional space/time continuum, therefore as such not space in the traditional meaning of the word.

would travel many, many, many times faster than the speed of light; so that one may place oneself throughout the galaxies within seconds. Thought takes no time to get anywhere - *thought is not in time.* Thought does not occupy space.

Q: It is not in any time continuum at all?

P'taah: Indeed. *Thought does not exist within the space/time continuum.*

Q: Thought is all energy.

P'taah: Indeed. But you see, beloved, in truth, whereas all of these scientific questions are very fascinating for humanity, the answers will not come until you know who you are.

Q: In that case, soul energy or the energy that is the soul - could you elaborate on that a little? How it is connected to the life-form?

P'taah: It IS the life-form. There is no separation between you and your soul. Without the soul there is no you. And there is no separation between you and any *other* being, seen or unseen, in *all the realms - in all dimensions.* And, dear one, *there is no separation between you and the divine Source.* When all of humanity truly knows this, then you will have *all* your answers.

Now, the technology which is not in the understanding of your scientists at this moment is truly the power of thought. What will be discovered and what will be common usage on this planet - as is on other planets - is crystalline technology; you may say it is in its prenatal stage now upon this planet. The power, the fuel if you like, of this technology *is* thought. We have spoken of this in these weeks and we elaborate later if it is needful, but truly, it is not of great importance at this time. It is only for interest. It is a demonstration to you about the importance that you *do* know who you are in the acceptance and acknowledgment, so that you may create change, that you may create love, where fear was. Crystalline technology is indeed a great magnifier and storer of energy. If the predominant emotion upon the planet within this technology is fear, then this fear shall be magnified and cast abroad, creating great chaos. If the predominant emotion is love, dear one, then with the use of these technologies the love will be magnified, creating harmony and light throughout the galaxies, do you understand? That is why it is so important at this time that humanity comes to a true bottom line knowing. The mechanics of it, dear one, is called GOD I AM. No separation, to know that all, *all that exists - seen and unseen - is the breath, the thought of the Source*. That is the bottom line.

Q: (F) P'taah, I understand that we all will come to know the God I AM; so when we get to that stage and when these changes come about, will we need any technology? Will we not be able to just come and go as we wish?

P'taah: Indeed, dear one, but there again: you will be still living in a physical universe, hm? Indeed, that which is your own creative power to come and go will be extraordinary, as it is within other races of people within your star systems; but it is not that you will suddenly turn into angels. You will still be in a physical body, albeit it be a little different from this, less dense. You will still be living upon your Earth. You will be creating wondrousness upon your planet. You will be travelling in your physical bodies with other physical equipment from one galaxy to another. Now, you may decide that you wish to travel within your consciousness, without your body, and you will be able to do it. In fact, you do it now - some of you at will and some in states of sleep etc. We have spoken of this when we spoke about your dreams. So it is a matter of choice, but you will have the ability to make the choice, do you understand?

Q: (F) P'taah, the less dense body you mentioned, will it still belong to the world of polarities? And I also wanted to ask if the world you are engaged in is a world of polarities.

P'taah: When you think of polarity, you are really thinking positive/negative, as in good/bad. But you see, positive/negative is not necessarily good/bad. It is merely a state of a physical universe. You see, dear one, when *you* say polarity, you are in judgment. But if you were a *scientist* and if you were looking, for instance, at somebody involved in electric engineering and he is saying positive/ negative, he is not judging that one is good and one is bad. It simply IS.

Q: Yes, I must contemplate that.

128

Q: (M) P'taah, when the time for changes affects us all, will it affect you as well?

P'taah: It will affect the galaxies, dear one. These changes are not merely confined to this planet. [It is] a different alignment of star systems; it is not only this planet. It is a coming into a different alignment of universes. It is because you are not really accustomed to thinking of there being anybody else but you and your planet. And certainly the coming alignment is very important in terms of the history of galaxies. Now, that which be energies such as I, who are speaking forth all over your planet in assistance to the coming changes, is because we care; because it is beneficial to many worlds that your transition be one of great beauty, that it be one of joy, that it be one of love, that there be absence of fear. This in itself brings great joy - to be here with you all. On many levels there is assistance. Outside of the space/time continuum, you know, all of this is viewed in a rather different fashion, with overview, so to speak. With a foreknowledge of how and what may be occurring. This is not to take away your sovereignty, because indeed, all of you may choose. But again, we say to you that you are used to viewing your planet as a solid mass, and that is how you view yourself, whereas it is not solid at all. There is room for millions and millions and millions of creations within the same space. There is not merely one Earth.

Q: (M) We have the idea that time is the same for all of us, but I have the feeling that it is different for each one of us; and that it has to do with the beliefs we have, which in actual fact alter the time, so that we are actually not in the same time at all.

P'taah: This is so. But it is very subtle. You do create in a fashion your own framework, but then, you see dear one, you are the central sun of your universe and your reality is different to everybody else's. There are certain guide-lines laid down within the morphogenic resonance within the whole of humanity, so that there is, if you like, an unconscious agreement of how it may be. But within that agreed broad spectrum reality each and everyone of you create your own variation. And of course, how you perceive anything is different from how anybody else perceives it.

Q: So anything I might experience today can trigger memories from childhood, so that I re-live and re-experience that event in this time, or through projection also experience an event that has yet to happen, so that I am almost being thrown into a different time zone.

P'taah: Indeed, that is why we say that in truth thought does not occupy space and time. In your thoughts you can be in your childhood; you can project a thought of what you perceive to be your old age...

Q: Let us say someone approaches me and I feel fear, then it can actually be an association with childhood which I am experiencing at that moment and the denial of that is the holding...

P'taah: Dear one, get out of your head. You know exactly. You have been through it in all of these months - you know exactly how it is. Whatever is fear from your childhood, you will indeed recreate and recreate and recreate, until you align it, *until you transmute the fear into love*. It does not matter what the 'story' is, it does not matter that your thought is cast back or forward or this way or that way - it is that you create a situation *to feel a feeling*. And where the feeling is fear, you have the knowing to be able to transmute the fear to love and so change your reality.

Q: That is happening - to some extent.

P'taah: But indeed, beloved.

We will have one more question and then sufficient unto the time.

Q: (F) P'taah, I just wanted confirmation: What you are explaining is sometimes a bit beyond our intellect and if we ask our more expanded consciousness for experiencing what is beyond the intellect - if we consciously ask for that, will we experience that...?

P'taah: But of course you will, dear one, and I will tell you: you will experience, whether or *not* you ask for it. However, you may also ask that it be brought forth perhaps in not such a traumatic fashion. You do not need great drama to learn, you know?! You may create the learning through laughter. *(P'taah tenderly strokes the lady's hair and kisses her gently on the forehead.)*

So, beloved jewels, hm. *(P'taah focuses on a certain gentleman, whose heart seems troubled:)* Dear one, you may speak with us whenever you wish within these days. *(Then turning toward a lady:)* And you also, dear one.

Q: Thank you.

P'taah: But it is not to thank - it is *my* honour, hm?

Q: (M) Do you have all the emotions toward us that we have? Like impatience, perhaps anger about how slow we are or something like that?

P'taah: Does it appear that I do, beloved? *(The audience chuckles.)*

Q: No.

P'taah: Do I display great impatience and anger with you all? What does your heart say?

Q: It says no. But sometimes I feel silly to ask questions, which to a human would be frustrating, because it has been asked before.

P'taah: Sweet one, you may ask *any* question, *whenever* you wish. And we will say this to you all: there is no question that you may ask which is considered to be too trivial, which is to be considered not worthy of an answer. You know, we do not consider you to be stupid because you do not know everything. We do not consider you to be naughty school children because you have not learned the lesson yet. Truly, all of this really is an appeasement of your intellect. And *we will do anything, that you may know the God you are.* It is because we love you. When you will cease the judgment of yourself, you will truly understand *how* we love you. *(Ever so tenderly - touching everyone's heart - P'taah adds:)* It is my desire for all of you, that you will love yourself as I love you.

(To the hosts:) Our thanks, dear ones.

So, dear ones, we bid you a very fond good evening and we are in anticipation to greet you again in the next week. Go forth in love. in light, in laughter, *being* in every now moment. Happy journeying, dear ones. Good evening.

Q: Good evening.

Chapter 8

EIGHTH TRANSMISSION.
Date: 16th of October, 1991.

P'taah: Good evening, dear ones.

Q: Good evening, P'taah.

P'taah: And so, indeed, you are here with quickened hearts, computer switched on, eh? *(Addressing a newcomer:)* Welcome, dear one, indeed. So, how are you? You are enjoying?

Q: Very much.

P'taah: A holiday, indeed. *(Addressing now a couple on holiday:)* And how are you holiday-makers. You are adventuring?

Q: Yes.

P'taah: But, of course. *(Turning toward a gentleman, who travelled a long distance to participate in this event:)* Beloved, you fell asleep, eh?

Q: Yes.

P'taah: Well, that is alright. You know, when you are asleep, your mind gets out of the way. So, indeed you are about to receive the knowledge without the impediment of intellect. However, beloved, that which you desire to bring forth, conscious communication - when your body is not in evidence - may indeed be so. You know, when you desire to manifest any thing in your life, and you know intellectually, that you have the power within you, you say: 'Well, now we will put it to the test.' So you send forth the desire and the moment you have a doubt you have undone all the good work. *Knowing* contains no doubt at all. If there is doubt, there is not the *knowing* - contemplate it.

Now, let us speak about your bodies. Let us speak about health. You know, when you are sitting in this room, with your hearts quickened with excitement, every cell in your body resonates to that excitement. The cellular structure of your body - every cell, *every* cell has its own consciousness, its own integrity, its own joyous impetus of creativity. In this fashion, diseasement, ill-health, is *not* something which you catch, as if it were a stray dog. *The diseasement of the body, indeed, is a reflection of the diseasement within your heart.* Now, when there is the intellectual knowing of this and the body is still not functioning correctly in full harmony and you say: 'Indeed, I do know diseasement of the body is diseasement of the heart', we would ask you to look at your fascination with ill-health. We would ask you to look at the fascination you have with your bodies - always, always desiring it to be better. So, what is that, beloveds? The desire for it to be better, to be in better function, better to look at - to be fatter or thinner, to be less old, to be more fit - as you are judging it, so you are creating, indeed, that which you desire to get rid of. Now, we speak to you very often about your intellectual knowledge and that, which is truly the knowing within you. Your body knows who you are. Your body is a reflection of how you are and what you invalidate, beloved ones - again - you certainly empower. What you are fascinated with, you draw to you. Do you know, that you have a great fascination with bodies which are not harmonious. You have a great investment in illness, as you have a great investment in pain. We are speaking of emotional pain. And so, when you are having inoculation against diseasement, you do so believing, that you do not live in a safe universe, that you will catch something, that it could be terminal. But dear ones, your whole lives are - in fact - terminal, are they not? In terms of your physical embodiments, at this time you do not carry them around for hundreds and hundreds and hundreds of years. In this fashion we could say it is all terminal. That is the grand illusion. So, it is indeed for you to look at the fascination you have in the parts of you, that you believe do not function as they should. You may come into the wise of it, of how it did occur in the first place what you were believing and what was your emotional state to create illness. When it continues, it may be beneficial to look at the fascination of it.

Medicine, your pills and potions, indeed, they are valid. The only reason they work is, because you believe they will. Your special diets, the only reason they work is, because you believe they will.

And that which is doctor, is certainly promoting the belief in sickness. Now, we are not saying: do not go to a doctor. It is very valid that you do, because within the morphogenic resonance of humanity is the belief that a doctor will make you better. So it is indeed to go to a doctor, if you desire it, but it is also behoving you well, to look at the inside story, and *you* are the inside story, always, always, always. Everything, beloved ones, *everything* comes back to you. It is *you* who are the central sun of your universe. It is *you* in truth who are sovereign beings, who live in free dominion, *if you choose*. Look at the fascination you have with your bodies, dear ones, and look at how often you use your bodies in spontaneous joy. Look also at how often it is, that you suppress the spontaneous expression of joy in your body. The obligatory exercise, the obligatory straining and stressing of the body to make it better does not truly benefit you all so well. If you are doing this wondrous regimen of exercising, however it may be, and you are doing it in joy, then it is wondrous. Your body is indeed also divine expression, however it is, beloved ones, it is wondrous indeed. If you desire to change [the body], you may, of course, change; but the change - as in every other aspect of who you are - will only, *only* occur with the embracement, *not* with invalidation, *not* with shame.

Well then, do we have questions?

Q: (M) You were saying, that we have investment in ill bodies, so where does this come from and how does it work?

P'taah: Now, there are as many scenarios, beloved, as there are people. We shall give you a hypothesis: Some persons may indeed hold on to illness, because they are afraid to function in what they would term a normal world. A person may hold on to illness, because it creates the loving attention, which they would otherwise - in their opinion - not be worthy of. Does this answer your question? There are many scenarios, dear one, I am sure that you could think of many yourself. Well, indeed.

Q: (M) I work with people who have great physical challenges. I can see that there may be a fascination for myself regarding their

disabilities; yet many of them seem to be functioning as well, may be even better, than most people function with full capacity.

P'taah: Now, dear one, there is certainly a difference with the humans, who are birthed into disability. There are certainly grand, grand lessons to be learned. It is also a co-creation.

Now, we are speaking about the beloved ones, with whom you are spending much time, who are malfunctioned of the brain. This truly is quite wondrous also. Let us be very clear first: that which is ill health is absolutely valid. We are speaking about your judgment of bodies, which are not one hundred percent. The people who have created this for themselves in this incarnation, with what seem to be impossible physical disabilities - those, who have been born with physical disabilities - *have chosen.* It is also a co-created learning situation for all people, who are in contact with them. You see, beloved ones, you do very often forget, that this life is only one facet, one experience of thousands and each time you take on physical form - in that creative expression of third density - you do so for different experiences, each time. And so, in this fashion, where there are people, who are totally incapacitated in physical form, how do you think it is, when they must be carried, when they must be supervised for every physical function? There are many people who think, that these bodies should be terminated, because 'of what use are they?' Dear one, it is not for you to be hurt at this, because you know it is so and we know that it makes your heart very heavy.

(The man begins to weep - he is overcome by his compassion for the ones spoken about, the ones he cares for.)

What we will say is, that within these bodies, so helpless and so incapacitated, therein is the soul here for the experience, and for the grand learning of all the people, who come into communication with them. Dear ones, every, *every* creature, every human, every plant and flower, every bird - every thing upon this beautiful planet - is indeed sacred, because all, all is divine expression. You see, because you - we say you, yet we are speaking truly of all of humanity - because you do not truly understand the God you are. Very often you find it difficult to recognize the Godhood in *all* things and because

136

you do not truly love who you are, you do not truly honour who you are. Very often you have difficulty in giving forth that love and honour to all, all things upon your planet and, indeed, to your planet itself. As you begin to blossom and to flower into the knowing of the God you are, so, indeed, the vibratory frequency of who you are, reaches out and touches every thing and in this fashion you create the change. In this fashion mankind goes forth, preparing the way for the changes to come. You see, it is so simple - love who you are. *There is nothing to do.* Not to judge your bodies, not to be fascinated by the disfunctional, but to be fascinated by the spontaneous joy of the now moment. That is all. *Being* in your own Godhood. — Questions?

Q: (F) I have been practicing the acknowledgment, aligning the judgment and feeling the feeling. When the energy is released from that and it travels up, does that release the diseasement that has been caused by not feeling it in the beginning?

P'taah: But of course, beloved, *that is the miracle of transmutation.*

Q: So, that all works.

P'taah: It all works, dear one. That who we are works very well. What a surprise, eh? You got it right, hm? But of course it works. How can it not work, when the heart is open, dear one? Every cell in your body resonates to the joy.

Q: It is wonderful.

P'taah: Indeed. So it should. We would be giving you a black mark, if it would not, eh?

Q: (M) The ultimate question is: why is anything, right? Why was there a need for a something in the first place?

P'taah: Dear one, each time you ask a question, what you are reflecting to yourself, is a question about what you perceive to be God. What you are is a definition of God.

Q: I believe because we cannot conceive the totality, that there is something more.

P'taah: That will do. There must be something, else you would not be here, hm?

Q: That is true.

137

P'taah: Indeed. We cannot fault you there. There is not fault in the logic. No fault in the intellect, eh? But you know, truly, this whole exercise - what is called 'life, the universe and everything', is not an exercise in logic. It is an exercise of the heart.

Q: In emotion.

P'taah: But of course, beloved. It is about feeling. And when you will truly allow yourself to feel without judgment, you will know God, because you will know in that instant, that there is no separation, and then you will understand that the answer is not about logic, the answer is about your heart.

Q: I will have to contemplate that.

P'taah: Indeed. It is very wonderful that you are here and that you are giving forth the questions that you may answer with your heart, and you will, beloved. You will. We think you are very clever to have brought yourself here. It is called a grand creation.

Q: (M) It was said, the other night, that nothing is too trivial to ask...

P'taah: Indeed, have you got a good trivial one to ask?

Q: Yes. Just before I left my home, I saw a cockroach on its back.
(Laughter.)

P'taah: Did you poison it? Was that why it was lying on its back?

Q: I have put some poison out. What should I...How do I...?

P'taah: Should you stomp on its head to put it out of its misery?

Q: I left it feeling, that I have no right to dictate its future, but then if I put poison out, I have already made a decision about its destiny.

P'taah: Indeed. You have answered your own question, eh? You know, in truth, if you are truly aligned with that which is called cockroach, you would say - please be gone, and they would.

Q: I have asked that before, and have spoken to the cockroaches in the house and said that I don't enjoy them being here.

(Sympathetic laughter among the audience.)

P'taah: Did they go? Not even for a moment? *(More laughter.)* You have to be specific, beloved.

Q: Maybe I am too wishy-washy. Should I not enjoy them enough to say: 'Well, stay please'?

P'taah: It is always your choice. But you see, the more you are fascinated with the cockroaches coming to you, the more they will be there. It is perfectly valid that you should put out the poison for them, but it is really not the ultimate solution. There is no judgment about that, dear one. If you say to somebody, who is not practising this 'madness' called New Age, and you say I am not poisoning my cockroaches, I am merely asking them to go, then it is possible they would take you away in one of those jackets with no arms, eh? But you see, it is so that whatever creatures are eating your garden and are infesting your home and eating your food and scaring you out of your wits, in truth, you do not need to resort to that which is extermination. You may indeed, in all respect and honour, ask that your unwanted visitors do go. Sometimes, like your very own visitors, they do not, but when you are very clear, and when you are not in judgment, then indeed, you will find that they will go.

Q: It is rather roundabout, though, because if you wish them to leave, you are judging that they are not desirable.

P'taah: Dear one, do not confuse judgment with discernment. To say that you would prefer that they were not here, is not to say that they are bad or they are wrong. It is a matter of preference, not judgment. There has been much confusion and people are running around believing that they are wrong, because they are in judgment and they cannot have a thought without thinking that they are judging. We are not talking about wishy-washy stupidity, dear ones, we are speaking about discernment. What is it you desire for yourself, what is it you desire for humanity and for your planet, not in judgment, but in honour and respect, in love, and certainly, with discernment, well always truth, with discernment. You understand?

Q: Yes.

Q: (F) P'taah, today, I found out that my house has been broken into while I'm away. I don't know if I have lost anything, because I am not there, but what I am asking is - what is the lesson? It has never

happened to me before and I am just wondering whether I'm invading others that this should have happened to me.

P'taah: It is not that you have been invading others, beloved. And it is not only you, who are involved.

Q: There is somebody looking after my house and I wondered whether she was very fearful of being responsible for it.

P'taah: That is so. Now, it is also a demonstration for you to consider that there is always the thought in your mind that somebody would come into your house and that you should take some method of protection, do you not?

Q: Not a very strong one. I don't have very strong doors, and I often go out and leave my back door open.

P'taah: Hm? but what do you do, in spite of the doors being open, to ensure that you do not get broken into?

Q: I put a sign, a Reiki[1] sign, over my house, and I didn't do that before I left.

P'taah: Indeed. How extraordinary, Beloved. hm?

Q: So, are you saying that it is better not to try and protect it in this way, but just to trust that..

P'taah: Dear one, we are not saying that you should or should not do anything. We are only saying to be aware *why* you do things and what you believe. What is the bottom line belief behind your actions, hm?

Now, we are going to have a break that you may refresh your bodies, and you may indeed be thinking of wonderfully tricky questions for when we return.

(P'taah returns after the break.)

P'taah: That which is your body, dear ones, this expression of who you are, responds indeed to the child-like spontaneity and joy in your moment by moment existence. What and who you are at this time is living in pain, yearning for love, desiring wholeness. Now, when you

---

[1] Reiki, a protective sign or signs originating in Buddhism and introduced to the West by a Japanese. Reiki may be likened in character to the Shamanistic medicines of native Americans.

are, in truth, whole within - with who you are - when you are indeed experiencing love for who you are, then you understand that there is no separation; then indeed, everything else follows. Always, always it comes back to the love of who you are - acceptance, acknowledgement. You are here by divine right. That which is good health, that which is abundance, that which is joy, love - it is yours by *divine right*. You deserve *everything*. Why? Because you are here. Because, indeed you are divine. There is nothing in the universe that you may not have. There is a *but*, and the *but* is: *To have all the wondrous things that you desire, you must know who you are and love every facet of it.* Have you noticed that there is a what is called 'common thread' in all of these grand discourses? *(Laughter.)* Vibrant good health is your right, if you believe that it is. That which is *not* vibrant good health is also a valid and a divine aspect of who you are, like every thing else. Questions.

Q: (F) P'taah, getting back to the discussion we had before about cockroaches: There are things that suck your blood, and bite and poison. There are plants that sting and grab and cut. How does discernment come into that? And then there are the March-flies.[2] (This remark causes sympathetic laughter.) I really wasn't going to let any of the other stuff affect me, and it does not, too much, but the March-flies have got to me.

P'taah: And to our woman. *(Laughter.)* Well beloved, you better love them.

Q: Would it help to write a letter?

(Shrieks of laughter, as this is what Jani King did to a rat living in her house - and the rat vacated the house.)

P'taah: *(Humorously)* Well it worked for the rat, so why would it not work for the March-flies? Now, there is also this to consider, dear one: They may sit upon your body, and you may nullify the effect of the bite. It is the same as when we were speaking before of rat poison, indeed? When you believe that something is going to bite you, to hurt you, to scratch you, to poison you, then indeed it will. It is no

---

[2] March-fly, a seasonal stinging fly prevalent in specific areas of North Queensland.

coincidence and there are people who will say: 'Well, what about when you do not know when something is poison or you do not know that they bite, etcetera'. Well indeed, like everything else, it is called lessons to be learned. It is called lessons to be learned about a safe universe. It is lessons to be learned about what attitude and belief you hold about that which is animal and that which is plant etcetera. It is lessons to be learned about the reaction of the body and what is called emotional statement, Indeed? Like everything, it is all multidimensional, eh? On one level the answers are all very simple, but like every thing else, it goes on and on and on, hm?

(P'taah walks over to a lady and kisses her on the brow.) P'taah: Dear one, do not worry about becoming dependant on that which is I. If we find that you become dependant, then we send you away. Is that a good deal?

Q: A very good deal.

P'taah: And do not judge how you learn your lessons, and what you are creating. Bless everything that you bring forth. hm? I am a terrible eavesdropper. However, dear one, I usually wait until I have been invited.

Q: You have been.

P'taah: But I know. Very occasionally I drop in where not invited and sometimes I am asked to leave. *(The audience chuckles, knowing P'taah is referring to Jani King.)* Hm? Indeed.

(Once more the gentleman, who is especially concerned about the cockroaches:)

Q: P'taah, at the risk of being corny, if I know who I really am, and I love myself and I ask the cockroaches to go, they go?

P'taah: Indeed.

Q: If I don't know who I really am, and I don't love myself and ask the cockroaches to go, they don't go!

P'taah: You do not even have to be, what is called, 'perfectly aligned', but you certainly have to know that when you ask them to go, they will.

142

Q: Ok, then it seems that there is a veil. I believe if I ask the cockroaches to go that they do go. I believe it but it does not work. There is some factor I have to add into it. Is it that I have to learn to love myself, and then all those things happen?

P'taah: Dear one, when you love who you are, then you will also love the cockroaches.

Q: And you can ask them from that space, and they go?

P'taah: Indeed. Now, you also may have a wonderful experiment with this: It is to take your consciousness into the consciousness of the cockroach; to become the cockroach. It is as with plants and trees in your garden; when you become as one with the consciousness, with the spirit, if you like, of that which is the flora and/or fauna and you give forth a blessing and encouragement to the plants, they will prosper, and when you give forth blessings to that which is cockroach and say that you would desire not to share your dwelling with them, that they should go forth and find a more hospitable dwelling elsewhere, they will go.

Q: It is not just coming from your head, you also have to feel it?

P'taah: Indeed, beloved, that is how you create the reality.

Q: So I have to bring it from my head down into my heart?

P'taah: Indeed.

Q: Well, I seem to have a lot of trouble with that.

P'taah: *(Humorously)* No.*(Laughter.)*

Q: Yes. Any pointers on that, please?

P'taah: Dear one, we have been giving you pointers on that for quite some time.

Q: Indeed.

P'taah: Dear one always, always, always we come to the same [thing]. hm? - I love you. - Not to *do* anything. You are having wondrous practice at *being*, and it is wondrous that you are within the house, and in this fashion, as we have told you before, about being gentle with who you are. To be allowing of yourself. And you know that when you are, you begin to flower. Is it not so?

Q: Yeah, and it is also more difficult than it sounds.

P'taah: NO. *(Laughter.)*

Q: Yes.

P'taah: You surprise me. Everybody else got it at once. *(Shrieks of laughter. P'taah adds gently:)* We have said to you, every thing is on line. You are here to make a great discovery. So be it. Does that answer your question about the cockroach, beloved? Very good.

Q: (F) I can see two sides in me. One is that I am very fascinated with the patterns of diseasement, be it in myself or others. The other side is that it really makes my heart sing when I feel or see myself or someone else releasing tensions or limiting patterns. I am fascinated, and I wonder which is the right attitude?

P'taah: It is called dichotomy, indeed? As your fascination with wholeness becomes more and more and more, so you shall reflect wholeness more and more within you and so you shall reflect outside of Self more and more and more. Do you understand? Health is not a matter of will. It is a matter of beingness. It is a matter of the heart. And that is alright, you know, to be fascinated, in terms of that which is the healing, and it is certainly a great joy to behold, to facilitate that which be the healing; and we are not suggesting not to do such. It is merely to reflect for oneself how the patterning is, which creates diseasement.

Q: (A very young lady.) P'taah, I would like to know the best way to dissolve fear. Do I put myself into that situation, and then thoroughly feel afraid? Does that kind of help anything, or does that attract what I am fearing? I find that...

P'taah: What do you think, beloved?

Q: I feel that I should not really put myself in that situation.

P'taah: Of what? Feeling fear?

Q: To feel that fear, but not do it.

P'taah: Doing is only an extension of feeling.

Q: So that means I can feel it and dissolve it without doing it?

P'taah: Feel it and embrace it.

Q: Embrace it?

P'taah: You do not *do* fear, you *feel* it. You cannot *do* fear. Now, let us be very clear about this. You are speaking about fearing a situation.

Q: Yes.

P'taah: And you are saying: 'Should one create the situation to feel the fear?'

Q: Yes. In order to dissolve it.

P'taah: Now indeed, so let us be very clear: you do not *do* fear, you *feel* fear. It is not necessary that you create the situation. By the acknowledgment and embracement without judgement you may transmute the fear without bringing forth the situation - [in physical reality] - which is feared. In fact, beloved ones, we would say that this would be a very harmonious way for you to do it. It is not necessary for you to create all of the drama in physical reality. You may embrace the emotion of it, and so change the fear into love; not by *doing* anything, but simply allowing it to be, without judgement, acknowledging and embracing it, knowing, that within the fear is the jewel.

Q: (F) Just to confirm that. If I'm afraid of something and if I try to feel that I should not be afraid, then I am invalidating the fear.

P'taah: Indeed.

Q: So, if I am afraid of something, I first of all have to acknowledge that I am afraid of it and it is all right to be afraid. We have this belief that we should not be afraid.

P'taah: Indeed. But it is to go to the *emotion* of the fear.

Q: So in actual fact we have attracted some things to us to actually embrace fear, not to reject fear.

P'taah: Beloved one, there are only two expressions in your universe. One is Love and one is Fear. Both are valid.

Q: We have to embrace both.

P'taah: Indeed.

Q: (M) Say we have a fear of snakes. Should I imagine that I love a snake? You say: 'Let's make it clear', but perhaps I do not want to hear, or something... can you give my heart a direction?

P'taah: Very well.

Q: Let's make it on snakes, say.

P'taah: Indeed. You are terrified of snakes. We are speaking again hypothetically. We are not speaking about anything, that would not disturb you greatly; we could imagine, say, a pathological terror - let us make it truly dramatic. Now, this can be approached intellectually. You say: 'I have a pathological fear of snakes' and you can go to all your therapies to discover the bottom line. And people do so with their fears, whatever they may be; and that is very good, to discover the why's' - this is intellect. In theory we would say: If you have a pathological fear of snakes, then what you will draw to you is nothing more sure than that, which will express this fear. But it need not be so. It is certainly to understand that the fear - whatever the fear is - is a valid expression. It is to bear in mind, that whatever you invalidate, you will empower, you will draw to you. Do you understand that? And so it is to look at this fear of snakes, to imagine the worst scenario. What would it be like, lying on your bed at night and to reach out and pick up your book and there beneath the book is a snake. Now, when you imagine this scenario, you create within your breast the feeling, the fear, and what you are going to do is to transmute the fear to ecstasy. So, what are the rules, beloved? Take responsibility - align the judgment. In this case the judgment about yourself for being so weak as to have the fear in the first place, the judgment you have about snakes. Then you take off your head, put it under your arm and feel the feeling.

Q: You feel the worst possible outcome?

P'taah: Indeed.

Q: Imagine what is happening and scream or whatever..?

P'taah: Dear one, that is called expression, it is not even necessary to [do that]. It is about feeling. It is not suppression; it is not expression. It is the neutral place between [both]. Because feeling, indeed is neutral energy. It is the judgement which creates the good or the bad of it. And as you feel the feeling without the judgement, so the energy - which is the feeling - is allowed to move. In this fashion you need not draw to you the drama in physical reality.

146

Q: Keep feeling the feeling until it is light?

P'taah: Indeed. It is called embracing it into light.

Q: So feel it until it is 'Why am I bothering to feel this', do you mean that?

P'taah: Hm. You will know when to stop it. *(Laughter.)*

Q: I will definitely try it.

P'taah: I will tell you this. If there was something that you fear, you may say that you desire a gentle experience, which will bring forth the emotion, that is if the imagination does not do the trick. It is not to *do*, it is to *allow*. That is what your world is about: ALLOWANCE. That is what is called feminine energy, *allowance*. It is not to *do*. That is called masculine energy, that which humanity has been operating from in your history. Now, it is not to negate masculine energy. It is to bring forth the feminine energy that it may be in balance, and you may say, that intellect and that which is emotion, is also called the positive/negative. It is not to judge. It is not that one is right and one is wrong; they are both equally valid. It is a natural polarity. It is the judgment which creates the imbalance. So it is allowance, allowance of the transmutation of fear. Allowance of the fear itself. Embracement - not to suppress it, not necessarily to express it, merely to allow it and embrace it into the light of who you are. Is it clearer?

Q: I realize you can never give a straight recipe for this, because it is not like cooking a cake. There has to be an understanding, it is not just 'follow these physical directions'. It has to be a true understanding. It has to come with your own wisdom.

P'taah: Indeed beloved. You know there is a very simple recipe: Know who you are, and *know that you are God.* All else, really, is a continuation of that, because if you do indeed love who you are, then you do indeed love all things. *If you do know, in truth, that you are the GOD I AM, then every thing that you see outside of yourself is a reflection of that Godhood.*

Q: How do you deal with it when the ego tries to show you that you are not perfect, or 'look at this bad side of you'?

P'taah: But dear one, it is to know that every facet of who you are is an expression of *the God you are. every, every facet of who you are is an expression of divinity*, else it would not *be*, beloved. There is no right or wrong. There simply IS.

Q: I am working on a preconceived idea of what perfect is. That is where the whole trouble is.

P'taah: Beloved, you are perfect.

Q: Even with my 'whatevers'? They are as perfect as Jesus was?

P'taah: Beloved, you *are* perfect.

Q: My ego is perfect, my everything is perfect?

P'taah: Dear one, you see, it is humanity's definition of 'perfect'. As we have said before, your definition of 'perfect' is a finished object, pristinely finished, but you see, there is nothing in the universe that is finished, else it would not *be*, not even God.

Q: This, now, is absolutely perfect, that I am hearing. How do I retain it? Write it in my eyelids when I am not seeing?

P'taah: I will tattoo it on your eyelids. Indeed, beloved, we will tattoo it on your heart. *(And softly:)* It is alright.

Q: (F) I recently had the experience of my friends bestowing their love and showing their love for me and I found it is a little more difficult to receive the love than to give it, but not as bad as it was last year. It is quite different to be open, to accept and receive it than to give it.

P'taah: Now, dear one, is that not a perfect example of that which is masculine/feminine energy. Do you understand?

Q: Yes I do, now.

P'taah: And so it may be that you are creating a wondrous dance within your universe in coming into the understanding that you are *all* things. That you are masculine *and* feminine, both. The giving of the love is what you may call masculine energy, and the receiving of love, reception, is feminine energy. For most of you it is easier to give than to receive.

Q: (M) P'taah, you refer to us sometimes as light beings.

P'taah: So you are. Getting lighter and lighter.

Q: Is there more to that?

P'taah: How much more would you like?

Q: I don't know. I ask.

P'taah: Now, indeed that is what is called a multidimensional question, because you are certainly light beings in your other-dimensional aspects. Now we would remind you that when we are speaking of 'higher vibrational frequency' of beings, we are not speaking of higher echelon. We are not speaking of hierarchy. We are speaking technically, scientifically, if you like. As the universe, as your world moves forward into fourth density, into a higher frequency, physical density becomes less dense. that which is of lesser density. Higher frequency becomes more and more light, so that in the transition of the Earth - in that higher frequency - every atom and molecule will have a lighter density and will indeed create light within it. That is not only for humans, but also for the flora and fauna and for your whole planet. As it changes resonance to the higher frequency, it becomes lighter and lighter. Does that answer your question, beloved?

Q: To a certain extent, but not fully.

P'taah: Indeed. That which is 'you are light beings now', is that we are speaking of your own multidimensional aspects, because you are all things at once. As the past, present and future in your terms only exist in this space/time continuum, outside of this space/time continuum it is occurring at the same time. So it is with the aspects of Self that you could call frequency. Now this is outside space/time as you understand it, but it is also in, if you like, a different direction. But what you are is omni-directional. You are all directions at once, as you are all taking part in all time frames at once. Now we are getting fully technical.

Q: I love it.

P'taah: Hm, I know. And so it is that your are involved in your past, present and future; you are involved with your aspects of multi-dimensionality in frequency; of lighter and lighter frequencies. You are involved also with what is called probable realities, probable selves and the probable selves of all other time/space continuums. That means you are very busy. It also means, in truth, that you should not concern yourself over-much about all of the other aspects, but simply to know that you *are*. And as you are of the intellectual understanding of the multi-dimensionality of humanity, so you will come into more and more understanding that there is no separation.

Q: P'taah, it seems there is no separation. Ok, I can follow this thought, more or less. However, in our physicality, right now in this third density we seem to be all separate. We are all separate. I look at all these people. We are all separate. It is so difficult to understand the Oneness.

P'taah: Beloved, what you are doing is preparing yourself for the transition, and you will have the moments where you truly will feel the Oneness.

Q: I am looking forward to that.

P'taah: Indeed, I know, I know. *(Very softly:)* So be it.

Q: (F) P'taah, in connection with that subject: One day I woke up with the remembrance where I came from. I had the feeling that I was aware that I had just come out of a place where there was no time or space, and I cannot explain it in words. I just had the feeling and I remember the feeling.

P'taah: Indeed that is very valid, beloved, because in truth, the vastness from whence you came is totally beyond time and space, in the vortex called I AM, called GOD, and that is where you are returning to. And yet, of course, you are all there - that is the paradox.

Q: (M) P'taah, to put more clarity to that: energy, is that what it all is?

P'taah: Indeed.

Q: So everything is a conversion of energy at that particular frequency.

P'taah: At any frequency.

Q: Even ourselves?

P'taah: Indeed.

Q: How do we get identity and emotion, and everything else we feel, from this ?

P'taah: It is simply an energy that is, at this time, immeasurable, unmeasurable. You see beloved, there are many, many forms of energy totally unknown to your physicists. You may say that God is energy. Everything is energy.

Q: (M) So our soul is energy?

P'taah: Yes, beloved.

Q: So, soul is an energy carrier and spirit is energy.

P'taah: Indeed.

Q: Different frequency?

P'taah: All is energy.

Q: What is Spirit, energy? Special energy?

P'taah: Indeed. It is all special energy because all, all is of God, the All That Is, the Source. That is why truly there is no separation. It is just that you have to have the names for it all. You have to put it all into its own little box.

Q: (M) That is what third density is all about.

P'taah: What is that, beloved? Are we asking a question or are we having a dissertation, beloved?

Q: Having a dissertation.

P'taah: Indeed, go for it.

Q: I think it would be rather more valid for me to ask something which I am more in earnest of the answer. This concerns dreams I've been having. It is always being imprisoned and attempting to get out of the prison.

P'taah: How extraordinary, beloved. (*The audience chuckles.*)

Q: And always never quite making it. I can understand that it might relate to the feeling of 'victim', but I cannot see that in every day life, which I suspect would be its source, it could relate to anything other than the separation of Self from SELF.

P'taah: So you want multi-dimensionality in your answer, dear one?

Q: Yes, yes.

P'taah: Do you truly ask me to believe there is nothing in your day-to-day life in which you are not a prisoner trying to break out?

Beloved! *(Laughter from the gentleman asking the question.)* Enough said then.

Q: Yes, I think I am too frightened of the answer.

P'taah: You know full well the answer. You are also prisoner of your body, and you wish to break out. You are prisoner of many things you would wish to break out from, beloved. I would suggest that you rapidly embrace being prisoner.

Q: (F) P'taah, this is really somebody else's question and she was too scared to ask it: Things are going along swimmingly and suddenly, whammo, a car accident. Just comments on how these things seem to come out of the blue.

P'taah: Wondrous co-creation, eh? Just when you think it's safe. Just when you think you've got it all together, or just when you think you know it all. Whatever it is in your life, you have a wondrous way of creating that which is to make you sit back and go 'oh-oh, what did I do this for?' And truly, for each and every one of you, what you are doing in your day-to-day life is merely creating situations that you may look at the feeling that you may look at what you manifest to show you that within the breath of each and every one of you there are always levels upon levels of your belief structures and your emotional re-actions. Each of you could ask about specific areas in your life, particularly about accidents and say 'what does it mean?'. There has been, in this week, more than one accident amongst you. It is for you to say: 'How does it feel? What am I creating with this accident?' There is no accident - you all know so. It is learning to peel away the layers in your understanding, to come into the knowing. And beloved ones, we would remind you that it is not necessarily so serious, this learning. You do not have to create dire circumstance. You may do it in a light hearted fashion. To have what is called 'not a serious' accident is indeed to be preferred to one which is extremely hazardous to the health. Do you understand what I am saying? Life was not meant to be serious. Well, if you want to be serious, it's alright with me. Serious is perfectly valid, beloved ones.

Q: (M) P'taah, a different subject; Antarctica - does this have any specific meaning for us on this plane at this time?

P'taah: Both of the poles have significance for your planet. We would defer this to one other session, beloved.

Q: That is fine with me.

P'taah: Is that all right? Because really, it is sufficient unto the time of this evening, and when we are speaking more of the changes to occur etcetera, we will address this.

Q: Thank you.

P'taah: Indeed. And so it is a time of journeying forth again. It is not truly 'good bye'. It is joy to see you and we will simply be in expectancy of our next time together.

It is to know that all is God, and so are you. That is all, truly. *(To one of the men in the audience:)* You have been quiet this evening, beloved. It is alright. Your presence is truly felt. We will see you again.

(To the hostess:) Thank you again, beloved woman. *(To the host who is operating the recording gear:)* Thank you, engineer. It is called 'stalwart job'. Dear ones, our thanks indeed. As always, you know, it is truly a joy to be with you and we are certainly already in excitement of being with you again very soon. Bless who you are, because who you are indeed is that reflection of radiance which casts forth light over the multiverses. And we truly are in awe of the brilliance, and soon you will know that *you are*, indeed, the GOD I AM. We are just asking you to exercise a little patience. I love you all. Indeed. Good evening.

Chapter 9

NINTH TRANSMISSION.
Date: 23rd of October, 1991.

P'taah: Good evening, dear ones.

Audience: Good evening, P'taah.

P'taah: How are you all this evening?

(The audience responds accordingly.)

P'taah: So it is that we have many new faces. Indeed, it is the ripeness for the quickening of the consciousness of that which be the humanity of this planet. *(Due to the fact that about twenty newcomers have joined this evening's session, P'taah sees a need to do some 'ground-work' first, so he recapitulates briefly:)*

Now, in these days we have been speaking with you about vibratory frequencies, which all of you are. What you are is energy. We have been speaking with you about higher vibration. We said to you that higher vibration has nothing to do with higher echelon, has nothing to do with hierarchy, and this evening, to begin with, we would wish to elaborate a little on this. Now, what your social structure is, is hierarchical, that is: With your government of this world and with your society, in fact in your social-economic structure, there is always hierarchy. Now, it takes on various forms in your society. There is the hierarchy of education, that of economics, there are the grand power games of those who hold positions of power in your government, who in fact mold your day-to-day life with the law. Then for those of you, indeed, who are involved in what is called New Age, there is also hierarchy, dear ones, for all of you - in your heart - imagine that you are not as advanced as many about you, that you do not know as much. Well, you see, truly in the area of your spirituality - and what you are is spiritual being, *all* of you - there is no hierarchy.

There are some who look at those such as I, indeed, as Guru, prophet, whether they be of human form or whether they are using what is called human telephone in this fashion. And very often there is a tendency to what is very akin to worship. But you see, beloved ones, I am as you, because we are all expressions of the All That Is. There is no rank. It may be that you feel that that which be I - or indeed anybody else, who is your teacher - should be placed upon a pedestal. It is not so, because I will tell you this: who you are is great learning for that which is I. What we are doing together, you and we, is a sharing of energy. Now it is true there are many such as I who come forth to assist in an expansion of consciousness, but beloved ones, you are not merely consciousness. You are, indeed, grand spiritual beings choosing to experience this dimension of reality, and in the state of morphogenic resonance - that is the collective consciousness of humanity - you have chosen at this time, and in your past experiences of this planet, forgetfulness of who you really are, that is all. This body and this beingness that you are at this time is truly only one facet of the grandeur of you. We are in understanding that you become lost. That you do not know what to do and indeed how to do it, and we have been saying forth that there is nothing to do, because what you are, at this time, is learning to become human BEING. *Being* in every now moment, in the fullness of the moment. That you indeed cast off what have been the chains of your past to build the brilliance of your future from every now moment. In this fashion there is no hierarchy, and what you are doing, at this time, is coming into the conscious knowing of your own power, of your own sovereignty, so that never again need you be dependant on another for your own emotional well-being; for your own learning. For indeed, within each and every breast there is the knowing of all universes. You see dear ones, always, always it comes back to the very same thing. As you come into *your* knowing, as you begin to flower, you understand that when you love who you are, when you honour who you are, you are loving and honouring Divinity - the All That Is, the Grand Source. In so doing, so you will reflect out from yourself that light of love, that divine spark which will change everything, that will resonate through the universes indeed. We have said to you before that one of the bottom-line situations with humanity is the fear of not being worthy. This is for each and every one of you to contemplate, the 'not worthy', because as you believe that you are not worthy, this is what you bring forth in your day to day life. Understand that, indeed, you are worthy of everything, because you *exist*, because there is no thing existing outside of what is called God, God/Goddess, All That Is. How can there be hierarchy, when everything, indeed, is imbued with the breath of God?

You are worthy because you *are*. Nobody is higher than who you are. When we speak of higher vibratory frequencies, we are speaking technologically, that is all. Instead of a hierarchy, going 'higher', when we are speaking of higher frequencies, you could imagine merely 'broader' antennae, which pick up the knowledge and energy from all of the galaxies, indeed from that which is the All That Is. So it is that the antennae become broader and broader. As they become broader, so the knowing becomes more, and as the knowing becomes more, so the antennae become broader. Do you understand? It can be likened unto a spiral. It is called energy, and within the spiral is the vortex and the vortex is the All That Is. It is truly very simple.

Very well now. This evening we will commence with questions earlier, because there are so many people here. We would give forth that whatever is in your heart to query, speak forth. There is nothing which is too trivial. If it is requiring of a different answer than is within the sphere of the operation of this evening, *(for the purpose of the book's continuity)* then we will refer it to another time. Do not feel that you have missed out, because if you require an answer, then we shall organize it that you get it. Now, dear one, you have query to put forth?

Q: (M) I have some questions which a friend of ours has left with me to ask. Just let me find my glasses, please. You have taken me unawares.

P'taah: (Very quizzically) No.

Q: Yes. Now the first one is: You have spoken that our bodies are designed to last hundreds of years. I think I would like that. What do I have to do to live that long?

P'taah: Well, one of the things that is required is, of course, to understand that what you call death is an illusion. That there is never an end to anything. Because you cast off your physical body does not mean that your consciousness is discarded as well. Now, as in all things, this question is multi-layered, multidimensional. The body was designed to last many hundred of years. For a long time in your pre-history it did so. You have in your historical records humanity living and reproducing for this length of time. Humanity, that is humanoids in other worlds also live to one thousand years. Now, let us talk about what you believe about your bodies: Well, you believe that by the time you are getting to your three score and ten it is all over. You speak in wonder of those of your people whose bodies exist for over one hundred years and more, but not much more; perhaps one hundred and twenty or thereabouts. Many of the people who, in their remote villages and areas, are not in communication so much, generation after generation, do not truly understand that the body is not 'supposed' to last that long. Those of you who live in what is now considered a fairly 'normal' situation on your planet, that is urban, not only believe that the body is not supposed to last for much more than eighty-odd years, but also live with extraordinary stress. You do not know what it is to be gentle with who you are and you create diseasement within your bodies, because you do not understand transmutation - that is to change pain to ecstasy. And so you make it quite difficult for yourselves to fulfil the body's potential. Your body has great integrity. As you believe about it, and about yourself, *so it is*. If you desire to live for many more years than is considered 'normal', then it is to look at your belief structure about your body and about reality; to understand that you may separate yourself from the collective consciousness and step forth. It is also to embrace what is called death in your terms. Because, dear ones, if you want to be alive because you are terrified of dying - guess what?

Q: P'taah, the second question: Does the continent of Australia have a special consciousness or energy? If so, what is it, and what is its purpose?

P'taah: Indeed. Now, of course it does have a special energy. We have spoken before about the morphogenic resonance. We will do a recap. Every atom and molecule has a collective consciousness; an energy which is consciously aware. It is linked to every like atom and molecule. As the atoms and molecules, for instance, in your bodies, collect to become cells, so the cells have their own resonance to every like cell. The cells bind together to become an organ of the body and so the organ has a memory and a knowing of every other like organ. As the cells and organs become a human body, so the body has a resonance with every other human body. It is the same within the unseen consciousness. When you are birthed into a family, so that family has a resonance, a morphogenic or collective resonance, a frequency. The family lives within a village or a town or a city and the collection of the village, town or city has *its* resonance. The city is within a country and the country has *its* own resonance, 'aware-ised' consciousness. Those who belong to a particular religion: it is that the religion has its own consciousness. Do you understand? So, that which is Australia indeed has its own resonance which is unique. And so does every other country. Now, of course, every collective resonance comes together in what is the resonance called humanity. That which is your Earth, your grand Goddess, has her own 'entity' [consciousness]. Everything that grows upon the Earth, every mineral, every rock, every tree and every flower is in symbiotic resonance with the Goddess.

Now, this country is indeed grand and wondrous spirit. Ancient, of course. The people of this country, who were here before the Europeans, had a wondrous relationship with the Goddess and indeed, a relationship with the stars and the people of the stars. That resonance is still with you. Do you know, for you it so much easier to come into the understanding of the human relationship with the Earth than it is for many, many peoples. Those of you who were not born here, but were birthed elsewhere and have chosen to come, why do you think it is? It is indeed that within the breast there is a yearning for this understanding. The greater soul of you, the greater part of you, the knowing within has brought you here, whether by design or 'accident' on your part, so that you can fulfil the yearning within your breast. This country will have a grand role to play in the time coming closer, which is the transition of the Earth. Those of you, who live in this[1] area of your country, are very clever indeed. For here indeed is a soft and gentle energy, here indeed you could sometimes almost believe in the grand forests of the time before the time of Atlantis and the fall therein. It is a gentle, gentle place. It is a place of great healing. There are many of the star-peoples who come to this area and have always come here. So, dear ones, congratulations. *(P'taah looks at the gent who asked the question:)*

Dear one, do you have a question to ask for yourself?

Q: Yes, I do. I hear the words many times. I have trouble with getting in touch with my feelings, the 'feeling the feeling'.

P'taah: *(Jesting)* Well, I am sure nobody here knows what you are talking about. It is not merely you, beloved, it is everybody here, because humanity has been programmed *not* to feel, because 'if you feel you will die'. So, it is not so uncommon. Now, you are asking for a recipe, beloved?

Q: Yes.

P'taah: *(Referring to a certain gent - the one desiring a recipe for dealing with cockroaches and snakes - P'taah remarks:)* Our beloved recipe-man is not here this evening - he also likes recipes, hm? Now, how do you feel the feeling? Well, dear one, you know, one of the first parts of the recipe is to *stop trying*. Does it sound familiar? Now, we understand that you have great trouble with *being* and not doing, hm? Is it called a panic situation, when there is nothing to do? *(Humorously:)* Nobody else finds this normal, eh? Very well, we shall give you a recipe, beloved, to get to the calm place. You have spoken forth about being told to sit under a tree, we believe?

Q: Yes.

P'taah: Well then, you shall do it. Every day. It need not be a tree, dear one; it also may be the ocean. It is very important for all of you to take time for your Self. We are not saying that you must go out and meditate. There are many people who say: 'But, P'taah, it is so boring'. We say, if it is boring, do not do it. We are wanting you to

---

[1] The greater area referred to is the tropical North of Queensland, Australia. The specific region in question is situated around Cairns and the Atherton Tablelands.

do what makes your heart sing, to be in joy, to be in laughter; but it does not mean that you should neglect yourself. *(Addressing the gent once more:)* You are to take at least one hour, every day. We do not ask you to sit and stop thinking. We ask you to go and concentrate on that which is leaf or a small animal, a sea-shell and we ask you to look at it, *(and very softly)* until you discover God. And we say to you that indeed, when you do know that you are looking at you, it will engender feeling, beloved and then you may embrace it and when you are embracing the good ones, it is easier to embrace the bad ones. It is to know that feeling is feeling, it is only energy; it is not good or bad. It is only the judgment which makes it so. *(Aware of the man's plans, to embark on a long journey the following day, P'taah adds:)* Now, you are going forth, and we would say to you that you may return. Everything is choice, dear one, and there is no wrong decision. There is no wrong choice, there only IS. Do you understand? So, indeed, we are already in anticipation of seeing you again. *(P'taah's voice becomes very soft, expressing much tenderness:)* I love you.

> **Q: (F) P'taah, you have talked to us about fear. During the last session you mentioned the word doubt. Is it the same as fear?**

P'taah: It is, however very often doubt is not so much fear. We were speaking in these terms, about how you may manifest anything you wish and it is merely to put forth the desire, embraced by the knowing, the emotion. In the knowing, that as you have put forth the thought and as you have embraced the emotion of what you desire, so it has already happened. Except when you doubt. Now, the doubt may not be so much fear as uncertainty about your own power; that is not necessarily fear. We would put it to you this way: You know that you are manifesting every day whatever it is that you desire. When you are creating disharmonious circumstances in your life, then it is to look at what you believe about yourself, about the world, about the people etcetera. We would ask you to take a little mental note of all of the things you bring forth merely from the casual thought. When you say to yourself: 'Goodness me, it would be so beautiful, if..' As you think it, you are in the emotion of the beingness of the situation, and then you cast it away; and 'good heavens', it occurs! Then you will say: 'Well, I was only saying a little while ago, that this is what I desire, and here it is - what good luck' or 'what a coincidence'. Well, is it neither luck nor coincidence, it is you who have created it without even understanding what you are doing. If you are taking note in this wondrous head, when these situations occur, then you will come into the feeling of how they do occur. Then you may manipulate your own reality as you will. For when you understand how it is, you will do it *without a doubt.* Doubt is only the belief in your inability to manipulate your own reality. So it is not necessarily a fear, but very often it can be. The doubt we were speaking of was not so much a fear. Do you understand? Has that answered your question?

Q: Yes it has, thank you.

Q: (M) I would like to know how one can generate lots of physical, emotional and psychological energy.

P'taah: Well, you know dear one, energy is not what you would call in short supply. In fact, there is an endless supply of it. It is a little like love. Now, again much of it concerns your belief structure about energy, about your own creativity, your belief of your own abundance. We mentioned before that it behoved you well to make a list of everything you desire. You can put one thing on one page each. You may draw a line down the middle and you may write your positive beliefs on one side and the negative beliefs on the other. You will find that much of what you believe in a negative sense are things you did not even know you believed. As you look at that belief structure, it is not to invalidate it, it is to understand that that belief does not serve you any longer. Then you may look at what you know intellectually to be true and you may embrace that emotionally and you will find that you will have endless supply of energy of whatever kind you desire. Does that help you, dear one?

Q: Yes, I think it will.

P'taah: Indeed. Now, beloved ones, we will have a break and we would ask you be quiet just for two minutes during that transition and we will be back very soon. While you refresh your bodies you may dream up some wondrous questions. Well, indeed, you are creating some beautiful light this evening. It is a joy to behold you.

Q: (M) Before you go, P'taah: How do you see us? Do you see our auras?

P'taah: But of course, beloved. That is only energy. You can see auras too, you know? In fact some in this room do. But, dear one, rest assured, I do see your physical body. Very beautiful it is too.

Q: Thank you.

P'taah: Do you know that you all, *all* have very beautiful physical bodies? All, all of you - very beautiful. *(Teasing:)* Not so beautiful as I, of course, but very beautiful. *(Chuckles in the audience.)* That is a tease for you. At one time, when we were with a wondrous group of people, I was saying that, indeed, if I were here in my own body, I would give you a light show to show you how beautiful you are; and one dear friend said: 'But dear P'taah, what would they say about the green scales?' *(Laughter)*. I said: 'It is alright, beloved, you cannot see the scales for the light'. *(More laughter)* And so it is with all your imagined scales, beloved ones, in truth you are radiant in your light.

(After the break.)

P'taah: And now, dear ones, let us continue with the journey. Questions.

Q: (F) Thank you, very well. Tonight is a blue Moon, and in many of our cultures the Moon has a special significance for the female or feminine energy. I was wondering if you could talk a little bit about that, please, apart from the normal scientific stuff that we know.

P'taah: Indeed, dear one. Well now, you know that in some of your cultures the Moon is masculine, hm? However, in many of your cultures the Sun is considered masculine, or what you would call positive energy, and this is very correct, because the Sun coming forth creates the flowering. It is virtually as if you would say the Sun is the germinator of the flora of your planet. However, the Moon would be considered by many cultures to be feminine energy - negative. Now, we will remind you, that when we say positive/ negative, it is not good or bad, merely opposite polarity. The Moon, indeed, is what you would say to be the soft side of the Sun. The Moon creates your tides. There are many cultures which practice planting by the Moon; calendar by Moon. So the Moon regulates the heartbeat of your planet. Indeed, you may say, in a quiet way, that which be feminine energy, whether it be within male gender or female gender, the intuitive, allowing, receptive energy is the heartbeat of humanity. The creative, intuitive. You see, dear ones, if it were not for these aspects, you would not be here, for it is these aspects which allow the flowering of humanity. Old cultures, old civilizations no longer on your planet, paid great honour to the Moon, to feminine energy. It is not to say that one is better than the other, because we come back to humanity *being* masculine/feminine energy, both. As we have said before, humanity has - for eons of time for a whole sacred cycle[2] - been operating on masculine energy. We are coming into a time, which is called the last cycle of this epoch, this era, and it means that it will come into balance. What humanity is learning at this time is to allow feminine energy to arise within the breast of each and every one of you; to allow that which is intuitive, receptive - that which, like your wondrous goddess the Moon, is in tune with your Earth, with what are natural cycles, because humanity of this time has forgotten. Humanity is not in tune with this goddess, Earth. So, as you understand about acceptance and acknowledgment of who you are, as you are coming more and more into the allowance, so that you may love who you are, every facet of who you are, so then are you coming into harmony with your planet. And as you come into harmony, and as your planet wroughts the changes to come, so the planet will be in harmony at a different vibratory frequency with the universes beyond your time and space at this time. Does that answer your question somewhat, dear one?

Q: Yes. You were speaking to John before about an hour's contemplation of whatever... I was just wondering if an hour's contemplation of the Moon ...?

P'taah: But of course, beloved. We would not recommend that you look into the Sun for one hour, *(laughter)* but to contemplate the

---

[2] Upon double-checking with P'taah in regard to the term sacred cycle, he stipulated its duration as 50,000 years, of which we are experiencing the 'last days'.

Moon in her silver beauty and to know it is a grand reflection of who you are. And for those of you dear ones, who are in male embodiment, it is to sit and contemplate this placid beauty to recognize the placid beauty within you.

Q: Thank you.

P'taah: Questions.

Q: (F) Are you perhaps..I may have heard of you or read of you as the patron of the arts and the crafts through the ages of time.

P'taah: Dear one, that which be I, in truth, have been called many things. Sometimes I have been called 'that old bastard'. That is our very colourful woman, but you know, as you are every facet of every human experience encapsulated in this little person, so this energy also *(referring to himself)* is as multidimensional as yours. We are not different, dear ones, we are merely other facets of who *you* are. We mirror to each other, always, the diversity and grandeur of divine expression, for that is who you are too.

Q: (F) P'taah, I was contemplating what you were saying about pain, resistance, energy. I was contemplating fluidity. I think you have mentioned it in several places. Science mentions about 'solid' matter not being solid but fluid, and I was thinking about our attitudes; that the less resistance we give to the attitudes, as our thoughts vibrate so our bodies will vibrate. We will become more fluid?

P'taah: Indeed.

Q: So that some of the things we fantasize about, like moving through matter or...

P'taah: Dear one, do you understand that you simply cannot imagine *anything* that is *not?* Hm? If you can imagine it, it IS. That is how fluid your universe is. Now, as you come into the desire, the imagination, as you put forth the thought embraced by the feeling, so your universe rearranges itself to accommodate you. Does that sound like great fluidity? Indeed it is. Your Earth, that you think of as a solid mass, is not. The floor you sit on is not truly solid in real terms. Now, matter is merely a lower frequency than that which you think of as thought or light. That is all. Your world *is* fluid. There is not only one

Earth any more than there is only one reality. And of course, you may say that the limited boxes of your understanding are what make it concrete. And as you become more and more unlimited, so the walls of the concrete boxes of your understanding crumble away to create vast, unlimited vistas. That is how you will, eventually, travel through the galaxies, indeed.

Q: (F) P'taah, with our shift in consciousness, is it going to take a long time for us to start seeing the lights around everybody; the lights around plants, the light itself? Will we eventually see it?

P'taah: But of course, beloved. Do you know how long it will take? The blink of an eye. I promise. Time is no-time, you know. Super-consciousness, that elusive ideal. The only thing that separates you from your SELF is your belief. Where there is no separation of Self from SELF, there is indeed the transition. It is like a gossamer veil, one little breath is all it takes. And some of you indeed have had a glimpse of what it may be, but you know, in truth, all of you *already know*, because all of you are *already there.* So you see, beloved ones, you are striving and scrabbling and struggling to be spiritual masters and you already are. That is the dichotomy. It is so simple. Love who you are. Know indeed that YOU ARE GOD. That is all. You are all grand spiritual beings *choosing* this dimension of reality. You are the thought of God. You are God smelling the rose. In the eons of time that you have incarnated in physical reality, you have been so caught up in the intensity of this sensual experience called life that you have forgotten who you are. But you know, it is only *you* who have forgotten. In the unseen realities there are millions, myriads of beings who have not forgotten who you are - who are waiting, indeed, for that which be third density humanity [to come] into the realms of 'no separation'. That you may have the knowing of your own God-hood and in that knowing, still experience in physical reality, although it will be a different reality, because it will be of higher vibratory frequency, where every atom and molecule gives forth the light of its being.

Q: (M) P'taah?

P'taah: Indeed, dear one. You are not being so quiet this time, eh?

Q: No, not this time. My question is: As opportunities come up in my life, I see two sides in me. One is fear and one is the desire to expand and grow. I find myself in difficulties to deal with both sides; to come to the right decision and to find the right attitude to know which way to go.

P'taah: Now, I find that very difficult to believe. *(Chuckles from the audience.)* So on one hand there is the excitement and the desire for expansion, for stepping forth, and on the other there is the fear of letting go. Hm? Does that describe it?

Q: It is not only the fear of letting go. It seems like the fear of death.

P'taah: Indeed, that is the ultimate letting-go, in your terms, eh?

Q: Yes.

P'taah: Well, nobody else here would understand that, hm? *(Laughter.)* Well, there are only two things in this third density: fear and love, hm? That is all. Everything that is not an expression of love is an expression of fear. Now, how do you make a decision? Well, we have said already that there is no such thing as a wrong decision. And if you choose fear, that is alright, because dear ones, that which you do not embrace you will bring forth again; time and time and time again, until you do. [The fact] that you recognize the situation means that you are already more than half way there. Is that a consoling [thought] for you? Hm. Now, we have given forth the answer to this many times, beloved, and we will do again, because it is very important that you understand how to deal with the fear.

Take responsibility. Align the judgement. Put your head under your arm and *feel the feeling.* It is called transmutation, hm? Now, you are in great judgment about fear; about your fear of anything, and you recognize this, because humanity is very good at hiding from fear. We are talking about the rationale. We have spoken before about this, that when society supports your fear, makes it 'alright' that you are terrified, then 'it is OK'. You think: 'Well, it is alright - I do not have to deal with it'. Of course it is not so. Fear is a valid and divine part of who you are. It is to be embraced. It is merely to understand that it *is* valid, else it would not *be*. It is to embrace it into the light of who you are and to always put forth the desire from your heart - to say that you desire from the God/Goddess of your being, that you may bring forth the situations in the most gentle fashion, so that you may come into embracement. For, dear ones, the only way you create change is with embracement. As you come into the embracement of the fear, each time you take a step, so you are creating more and more and more expansion. It is understood that very often you find it difficult to step forth from limited belief structures, the concrete boxes of your understanding, to leap into the void, not knowing what is called 'outcome'. It is, because you do not understand that you *truly* live in a safe universe; abundant, benign, joyously creative, of great integrity. And who you are in truth is also of great joy, of great integrity. Every cell of your body is creative, powerful, of great integrity of SELF. It is for you to remind yourself that truly it is a safe universe, and that, indeed, what you desire may be brought forth in harmony, for gentle learning. You are always safe, beloved ones, *always. Nothing* can harm you.

Q: So what is trust in this process? Is it trust that everything is safe?

P'taah: Beloved, what about trust in SELF?

Q: Yes, what about it? (Laughter)

P'taah: Hm! Don't you think it is about time for it?

Q: At times I am not sure what to trust, and what it really is.

P'taah: Well beloved, I will tell you this: You got this far, hm? What is the worst that can happen? That you will die? Annihilation, eh? Well you know, it is easy to intellectualize about death merely being a transitory state and *(touching the head of a very young lady)* for some it seems a very long way off. But you know, fear of death you have known lifetime after lifetime after lifetime. It is very ingrained in your morphogenic resonance. But you may change it, and as you learn to trust who you are and as you come more and more into the *feeling* of being in a safe universe, so the trust, the act of faith, if you like - to leap for love instead of lying down for fear - will become easier and easier. There is nothing that is not transmutable. And indeed, it is to know that there is no such thing as wrong choice. Is that not a relief?

Q: That is beautiful.

P'taah: Indeed, dear one.

Q: (M) As we all go through the process of expanding our consciousness, what happens to our future incarnations as to where and when and what form of body we will have?

P'taah: Well dear one, these are many questions all rolled up in one. We will start off with the future incarnations. Well you see, there is in truth no future. There is only now. Outside of this space/time continuum, that which you regard as past and future and present, is all occurring at the same time. As you come into the transition, so all, all will change. As you come into an understanding of what transmutation of fear is in every now moment, so you are already creating the change in what you call your past lives and what you would call future lives. You see, dear one, there is no separation, in truth. There simply IS. And so, when humanity, and the flora and fauna and the whole of your planet step forth into fourth density, everything that you understand as past and future lives, goes with you. Indeed.

Q: (M) Would you explain, please, what you mean by fourth density?

P'taah: Indeed. Truly, it is just a name. In that which is called 'New Age', everybody is very busy giving numbers to densities. And as the only density you know is this one, it does not matter what you call it. You could call it four hundred and sixty-fourth density, really. But what it means in, if you like, scientific terms, is that the Earth and the humanities therein are coming into a cyclic change, which is what all this 'New Age' stuff is all about, you know. Where the impetus, the earnest desire, the fervent yearning for change, is creating it. You see it is like a giant circle, you want it and it is going to happen. You get it. And as this change occurs, the vibratory frequency - and that is what you all are, energy is a vibration, a frequency - the energy, will vibrate at a higher rate. Higher and higher and higher. Your scientists believe that light is the highest density. Matter is the heaviest density;

that which you call a rock, or your Earth, is of heavy density. The difference between pure thought, which in your physical terms is the highest frequency, and the lowest frequency, is consciousness. It is consciousness which creates matter, which changes matter, also. So in this time to come, *which is very soon in your historical times,* the Earth will shift and change; the consciousness of humanity will expand and expand and it will all rise to what is called a higher density, that is that *the human body will be lighter, and it will create light. Light bodies. The Earth herself will be bathed in light, because as I have said before, every atom and molecule will give forth the light of its beingness.* And so, when we speak about fourth density, we are speaking about coming into that transition. Indeed, dear one.

Q: P'taah, so if I understand correctly, there is really nothing to worry about, then. I mean, ultimately everything is quite beautiful.

P'taah: Not even ultimately. We find it quite beautiful right now.

Q: Fair enough. I accept that.

P'taah: Because you see, dear one, there is *no future.* There is no future, there is no point in waiting for it all to get better. It will not. There is no point in saying you will be happy 'when'. There is no future, beloved ones. There is only, ONLY NOW. And it is of every now moment that you are building your tomorrows, and if you are building your tomorrows on fear and trepidation of the changes to come, what do you think you will manifest? What do you think you will bring forth?

Q: Pain.

P'taah: Indeed. And so, it is to be without fear in the knowing that what you have is exquisite beauty at this moment. Who you are is exquisitely beautiful. Who you are, indeed, is divine expression. Every facet of who you are. Every thought. It is NOW, beloved ones. It is NOW. There is nothing else. It is to embrace the shackles of your past into the present, to create your tomorrows in joy and laughter, in the knowing that you live in a safe universe; in the knowing that nothing, *nothing* can harm you. You are sovereign beings.

Q: (F) P'taah: If we have created a diseasement through a judgment, and we do not know what judgment we have made that

created the diseasement - although we ask for the knowledge - how do we dissolve the judgment, if we don't know what judgment we have made?

P'taah: You may love indeed the diseasement. You may desire from the God/Goddess of your being that the knowledge may be made manifest and it will be. You see, nothing is cast in stone and it is not necessary to consciously identify every emotion which has gone into creating the diseasement. For I will tell you, it is not just one [emotion], it is many. Very often we speak in very simplified terms about this, but your pain - which is always your pain of invalidation - is not created from one circumstance; it is created from many, many circumstances. Each building upon the other, until humanity is so lost in the morass of pain that it is impossible to unravel the threads. It is not always so, but it is very often so, and that is alright. You see, you have great judgment about diseasement and you are all desperate to get rid of it. And what you invalidate you empower. Very tricky stuff, eh? It is not necessary to identify every thread of the skein. It is merely to acknowledge that diseasement of the embodiment has been brought forth by pain and anguish, but it is not to invalidate dis-ease. Disease is indeed valid, it is also divine expression. It is to be embraced in the knowing that, of course, within it is that pearl of wisdom. You will not uncover the jewel in the lotus until you have embraced it. In this fashion you may effect the healing. It is not necessary to go digging around to try to discover every little nitty-gritty reason for this or that; it is simply to acknowledge that it is, and that it is valid. And as you come into loving who you are, and being in joy and peace and harmony with who you are, then all things will change.

Q: (F) P'taah, you said that star-people come to this area. Can we..

P'taah: Ah, we were waiting for this one. *(Chuckles in the audience.)*

Q: Can we make contact with them and how do we do that?

P'taah: Hm, it is possible, it is probable, all things are possible and probable. Why do you think that you have not already [made contact], beloved?

Q: I cannot remember. Consciously I do not think I have.

P'taah: Very well. *(Jesting:)* That is why you have me, eh? Now, every thing will happen in the ripeness of time. *(Directing his attention to the man at the lady's side, P'taah remarks:)* Indeed, dear one?

Q: Sounds familiar.

P'taah: It is interesting that you two are sitting together. *(Members of the audience chuckle, as they know both persons as being rather impatiently waiting for the global changes to occur.)*

Q: Co-creation, I presume.

P'taah: Both of you are yearning to be carted off in a beam-ship[3]? Well, most of you here are also [yearning]. And that is alright, it is good fun for you to contemplate this. I will tell you this though: If it were happening in your reality, there might be a little bit of terror to transmute. Hm, all in the ripeness of time, beloveds. You know, *the time is coming very soon, when the communication - face to face - with your brothers and sisters from other worlds will become very common place. They are waiting for you.*

Dear ones, sufficient unto the time this evening. *(P'taah addresses a certain gent, who had to travel a long way to come to this event, and had fallen asleep during the session:)* So, dear one, you are going to give it another go next week.

Q: I am.

P'taah: Very good. Do not judge yourself when you go to sleep. It is perfectly alright and you know, truly, you do not miss anything, whether you are here in your physical body or not.

(As usual, P'taah thanks the hosts:) Thank you, indeed, dear one. Beloved woman, our thanks. Our beautiful host and hostess, to allow us all to gather here in this wondrous and peaceful place. Do you know, as you go forth - each and every one of you - after this

---

[3] Beam-ship, commuter-craft between the Pleiadean mother-ship and Earth.

communion, you take the peace with you. You take the lightness of your heart and as you come into communion with your friends elsewhere, so you give forth the energy that is the feeling of what it is right now. And you know, when you come into a space within yourself where it is not so harmonious, take a little time - when you sit beneath the Moon or under a tree - and remember how it feels; and as you bring forth the remembrance of this emotion of joy, of tranquillity and harmony, of the lightness of your heart, so the disease which has been within your breast will be embraced, dear ones. It will be embraced into the light of who you are and who you are is grand and glorious; and we love you. Just remember, beloved ones - none of this is meant to be serious. Go forth with joy and laughter, have fun, be light in your heart as you were as a child, full of wonderment and give thanks forth to your beautiful Moon and your beautiful planet. Know, that in truth it is really a reflection of who you are. Good evening.

Chapter 10

TENTH TRANSMISSION.

Date: 30th of October, 1991.

P'taah: Good evening, dear ones.

Audience: Good evening, P'taah.

P'taah: *(Addressing a certain gent.)* Good evening, magician. Well come, indeed. How are you finding this grand northern area of your continent?

Q: Wonderful.

P'taah: But of course. We have spoken before about this area of your country and, indeed, you are very blessed. You are extremely clever for bringing yourself to this place. This wondrous place is of grand feminine energy, a place of the Goddess, indeed, a place of healing, of nurturing - so it is, that you are very clever. Now, that who you be, grand masters indeed, that you created yourselves in this place, this northern area of your continent, is what may be called a demonstration of higher energy desiring, yearning to come into the knowing and balance of human *being.* We said before that you create your own reality. It is simply the thought, embraced by the desire, that goes forth and the universe will rearrange itself to accommodate you. You are indeed most powerful. *You are not victims, unless you believe that you are;* and the foundation structures of your reality are your beliefs.

Now, as you come forth, incarnation after incarnation, you bring with you an overall game plan. That means what you shall accomplish is to learn to bring your experiences into harmony and each time, within that larger framework, you create your reality moment by moment. That is why we have said to you that in truth there is no future, *there is only now.*

There is only the now moment, because it is from the fullness of the every now moment, that you create your next now moment. When you hang on to the shackles of your past, when you are living in your past and *when you desire an outcome of your future*, you are forgetting entirely about now, your present. So it is, that you feel victimized, you feel out of control with the circumstances in your life.

At this time of great change, the cyclic changes of your planet, and indeed the galaxies and the change in the consciousness of humanity, you may create the transition in the easiest possible fashion. You need not be sitting about mulling over your past and waiting for dire circumstances to strike you tomorrow. Those of you who are here, are here because the area itself is of Goddess energy, it is to help you, if you so desire it - and indeed, you do at soul level - to come into the flowering of feminine energy, that the transition may be an easy one. For, dear ones, what the transition is about, is balance. In this period of your history humanity has been operating - both male and female gender - from masculine energy. Now it is time for the Goddess energy to come forth without fear in male and female, that the whole of humanity may be receptive, allowing, intuitive, creative - balanced with the intellect. So you are learning about how to expand feminine quality. Those of you who are female very often believe that what you are is indeed feminine. And, of course, dear ones, you are looking wondrously feminine; but you see, the history of woman - in this period of time in your history - has been one of struggle against the subjugation and where there is struggle for survival it is called masculine energy. You will know, very recently there was an upsurge called woman's liberation. The only thing that was liberated was what is called stringent law - to be relaxed. You see, those dear ones who have struggled for equality have been struggling for masculine power. Those of masculine gender have been in terror of feminine power. In your culture, that which has been of feminine energy has been crushed, it has been, as you all know, the dark side of the Moon. Man has been suffering of broken heart, born of guilt, because of repression and suppression of feminine energy within them. Allowance, hm? Allowance. And so in this time of change, certainly this place, your country and your continent, which has a reputation for non-allowance of femininity, may take the initiative and be a forerunner. For indeed, to be the forerunner is a probable reality for this country.

As we have said before, you have incarnated, all of you, on the planet of this time, to bear witness to the changes and to assist. And we have read the morphogenic energy of this country, to know that in many fashions Australians feel left behind. Beloved ones, it is not so. Not only are you not left behind, but indeed, there is capability to step forth and light up your planet. Each and everyone of you here, in kindling the flame within your breast, the yearning, the earnest desire for the light, is creating that which you desire. But you see, it all starts with you. Each and everyone of you singly! So beloved ones, we come back to the same boring old story: that it is you, that you indeed are the central sun of your universe and as you come into the allowing of every facet of who you are - to be embraced into the light, into the light of who you are - so indeed you will accomplish your soul's desire.

Now, we will start with questions at this time. Again, there are many of you, and if there are no questions, well, we would say 'congratulations' and we shall leave soon. Ah, indeed, our technocrat, eh? *(Laughter)* How are you this evening, dear one?

Q: (M) Fine, thanks, P'taah.

P'taah: Very good.

Q: P'taah, you were saying that Australia, and particularly the North here, is a safe place. Could you elaborate more on that? Is it safe physically, or is it because more of this energy is applied to this area?

P'taah: Dear one, there is no place which is not safe. You do live in a safe universe, whether or not you know it. So let us be very clear about that. In your recent years there have been grand scare tactics, telling you that you are not safe. However, I will tell you this: Is it Earth changes that you are referring to, beloved?

Q: That is true.

P'taah: Earth changes called earthquake, called tidal wave, volcanic eruptions, have already started upon your planet. It is certainly not something which is terrifying, it is a cause for celebration, it is the Goddess stretching, renewing, preparing. So, you would say: 'What about the humans who have lost their lives in these situations and what about the floods where there is farming, and what about the diseasements?' Well, there are many humans on your planet who believe they do not live in a safe universe, and they are here to experience for themselves upheavals. But if you truly know within your breast that who you are is a powerful entity, then you know that there is no such thing as death - it is an illusion. There are many of you who are terrified that you will die before the 'big bang' and [cannot] experience the reality of fourth dimension. Well, you will not miss out. *I promise you.*

There is no such thing as an unsafe place. Now, you may find, that if there is some variety of upheaval in an area where you are living or visiting, that if you have within you the balance, the joy of the changes, the knowing of safety, the knowing that life is more than survival, then indeed, you will find that you will simply not be there, whatever grand cataclysm may occur. It is so, you know. Cataclysm, in truth, is also a grand opportunity, it is called bearing witness to the Goddess, it is a reflection of the Goddess within each and everyone of you. For, dear ones, *you are God/Goddess*, reflections of the All That Is. As much as you intellectually decry 'old God' with his beard up there, for most of you in this culture there has always been the denial of that which is the Goddess energy, feminine energy, the balanced part of the Source, the All That Is. As you come into more and more allowing of who you are, then so the Goddess arises. Questions?

Q: (F) P'taah, with these changes coming now, how is the consciousness of this group going to help in the change?

P'taah: Indeed, dear one, you are not isolated and any grouping of people, coming together for whatever reason, creates energy bands, which go forth to the farthest galaxy. You are not separate. Of course, beloveds, you are assisting in the change, that is what you have all desired. You see, it is very difficult for humanity to truly understand that you are not isolated. Indeed, you have a physical body which is separate from every other physical body, in spite of much hurrying-scurrying to join the physical bodies together *(the audience chuckles)*. You do not finish where your body finishes. You do not finish - there is no end to you. It would behove you well to think of this sometimes, dear ones. You may touch yourself on the body, the extremities particularly - like feet and hands - and you may say: 'I do not finish here.' And as you do so, you may raise your hands outward and upward and wider, to symbolize that who you are touches everyone, every plant, every creature, every human and many who are not - whether or not they have physical bodies themselves. Know that who you are goes through every dimension of reality and probable realities into infinity. You truly are most powerful, you know. Beautiful, too.

Q: (M) How do our space brothers fit into this picture, with the changes to come?

P'taah: Hm, dear one, everybody is to experience their own changes. But you see, we would not wish to have you think that there has been *no* change during the eons of your time, because everything in your reality changes at every moment. The only reason there is what is called a bit of a hoo-ha about this change is, that most of you believe it will be cataclysmic, and many of humanity truly believe that there is no thing left after; that humanity will destroy themselves and their planet. Well, it is not to occur like this. You have chosen. That which are the star people are having their own expansion and changes, but you see, for many of them - but not all - the changes are already known. For many of the beings outside of this space-time continuum, it is indeed the knowing of what they have already created, and what you are creating in their various probable realities. We understand that this gets beyond simple explanations. However dear ones, as much as we do desire to keep this very simple for your understanding, so that our words come within the boxes of your conscious knowing, we are also speaking to the grander part of you who already knows. We have spoken before about 'no time', and about probable realities. So we will not go into too much detail at this time. But you see, there are those beings who indeed have understanding of all of it in their own consciousness, hm?

Q: (M) Good evening, P'taah. Is it true, following on from what you said about the planet Earth - manifesting a feminine principle in terms of energy - that the planet Earth is a feminine soul principle?

P'taah: Indeed.

Q: And that the Father Sun would be a male principle?

P'taah: Indeed.

Q: Then it seems that the primary manifestation for the female principle on the planet is to create?

P'taah: Indeed.

Q: In our solar system then, there would be a balance with male/ female energies.

P'taah: That is to occur, dear one. That is part of the changes coming forth. And that is how humanity will come into expanded awareness - expanded consciousness - that they may tap at will into the Goddess energy, female energy; the feminine principle - negative. And so it is that all of humanity, whatever their gender, are to come into that allowance, so that there may be that balance. Do you understand? And it is the balance of those polarities that creates what is called a jump in frequency.

Q: (M) P'taah, would it be correct to say that the balance of male and female energy, when they are in balance, would create more happiness than what is happening now?

P'taah: Dear one, there could hardly be less. *(Rueful laughter from the audience.)*

Q: Well that's good. The other question is: Is it correct to say that we as beings are really eternal beings, without beginning or end?

P'taah: But of course, beloved. You are reflections of divinity, which is infinite.

Q: Thank you.

Q: (M) P'taah, is there anything on the practical level we could do to assist these changes which are about to come?

P'taah: Indeed, beloved. Love who you are. That is all that is required of any of you. *(Very softly:)* Love who you are. I do not know why you find it so difficult. *I* love who you are.

Q: (F) P'taah, just to refer back to the cataclysms, and also our parallel realities: the actual fact that we are here now, that we have no past or future proves that we have actually lived through the cataclysm - if you know what I mean?

P'taah: Indeed, dear one.

Q: (M) You refer to either negative or feminine energies. Is there a thing called neutral energy, that is neither masculine or feminine?

P'taah: Dear one, that is called perfect balance.

(Another gent, a newcomer to the event:)

Q: P'taah, thank you for bringing this to Kuranda¹. There is more information coming up now that January eleventh 1992 is a major spiritual graduation for the family of humanity. From your perspective, can you see that this is an accurate projection and should we support it in a similar manner?

P'taah: Indeed, but it is truly not such a big deal. It is indeed the beginning of the last cycle, and it is the date to light, having harmonic frequency which is as you would say: creating an impetus for expansion; for the raising of frequency. However, we would certainly say: celebrate this. We would say celebrate every day, and certainly, there will be many of your people gathered at this time to give forth energy for the changes. And we would say that in itself is wondrous and by all things it would be wondrous to take part in this, hm? However, dear ones, it is also for you to be aware that there is a lot of what is called hocus-pocus and everybody will say 'but how do I know?' [what is true]; we understand that there are certain people saying, if you do not take part in this conversion of energy you are doomed never to 'make the grade'. We would say that this is grand exaggeration, perhaps wishful thinking. You know, scare tactics to bring humanity into a leap of consciousness really negates its own principle. Do you understand? The way that you know whether a thing be a truth for you, is how you feel. That is the barometer. Hm? How does it feel to you? And when the words come forth - dear one,

1 The meetings with P'taah took place approximately 8 km west of the village of Kuranda, which is an Aborigine name for 'meeting place of the Spirits'.

when *my* words come forth - and you say: 'Just a moment, this does not feel like *truth* in my heart' then indeed beloved, that is the truth; [the fact] that it is *not* truth in your heart. We have spoken forth about judgment, and we have spoken forth about discernment. You will always know what is truth. *(At this point P'taah calls for a break.)*

(After a thirty minute respite, the audience reassembles.)

P'taah: And so dear ones, to continue. Questions!

Q: (M) P'taah, I asked you once before about the significance of the Poles in relation to the changes to come.

P'taah: Indeed, and you think it is the 'ripeness of time', beloved? Hm? Very well. We would not wish to keep you in anticipation. Now, always, always there are many answers to the question. So, we will chat a bit about this and that in regard to your Poles. Arctic and Antarctic have been in your history of time a dimensional doorway. Magnetic pole. It is not just one single point. It is like a portal for the star people and the people of your inner Earth, who have always maintained contact and have travelled inter-galactically using these gateways. In the changes the gateways are shifting and changing as the electromagnetic energy of your planet is shifting and changing. It is that - not only for your Poles, but for other areas on your planet as well - the portals, the gateways, are relocating in a physical sense. They are also becoming broader and broader as the consciousness of humanity becomes broader and broader. During your history there have been people who have travelled through these portals into other dimensions of space and time. And so what will occur is that which has been a very narrow area in physical terms will become broader and broader, although the portals are not only physical. You may say that the key to the doorways rests within you. There was the Bermuda Triangle. It is no longer operational. However, what will be occurring in the next twenty of your years is that the portal will grow until it encircles your Earth, thus signifying the change. We would also say at this time that your consciousness, in truth, like the portals, does not occupy a space and is not limited to a time. It is beyond space and time, and as we have said before, your planet, that you believe is a concrete solid mass, in fact is many Earths occupying the same space, but in different time continua. That which is called time travel within these portals, which have been fictionalized accounts[2], is in fact what you would call a hit on the truth, because you may use these portals with your own conscious energy to travel in what you perceive to be time. *(Teasing:)* Not tonight, dear ones.

Q: What a pity.

P'taah: Hm, but there again, if you truly choose it, so be it.

Q: (F) P'taah, I have this picture that by clearing our emotional bodies we are also helping to clear the emotional body of our planet Earth.

P'taah: But of course beloved.

Q: Would you expand on this a bit, and how we are part of the planet?

P'taah: It is indeed that nothing is separate. And that everything that you perceive to be outside of yourself is but a mirror, a reflection of who you are, and when you look at what is called the chaos, what is called the devastation upon your planet, it is indeed reflecting the devastation within humanity. *As you come more and more and more into the acceptance, the validation, the love of who you are, so everything outside of you in your perception, will also change.* As you can accept and embrace every facet of who you are as divine expression, you may look into the mirror and know that you are looking at the GOD I AM, as you may fall in love with who you are, so indeed your planet will reflect it. So, indeed the planet will come into the flowering unknown since the time of Lemuria[3]

Q: (F) P'taah, more and more information is being revealed about working with electromagnetic energy, time travel, about a lot of the things that have happened on the planet since the last war. For example the Philadelphia Experiment. Now people are working with these principles on the external, using machines. As we learn, can we do this without machines, just with our consciousness?

2 Science Fiction literature.
3 Lemuria, an ancient continent before the time of Atlantis in the Pacific and Indian ocean region, of which the continent of Australia is a remnant.

P'taah: Indeed. Now, there will evolve a different kind of machinery, crystalline technology, which will be coming forth very soon in your time. Yet, as we have said before, crystalline technology is also a great magnifier. So, if these machines, which are powered by thought, are released into your atmosphere, that which is propelled by fear - and we would say that every thing that is not love is fear, including lust for power, which is merely the fear of being powerless - as these energies are released upon your planet, they would be magnified beyond your wildest dreams, so it may wreak devastation. It is preferable that this technology is put on hold, if you like, until the conscious energy of humanity is resonating to love, to the light. Thusly, as this crystalline technology creates the magnification, not only across this planet, but across the multiverses, so the magnification will be one of joy, of light, of love, of laughter. Your scientists have been playing with this one for many years and there is much secret knowledge among your scientific community. There have been many of your scientists who have been in contact and are still in contact with the star people. *There has never been a time when humanity has not been in contact with the star people and now, indeed, many of the people on your planet are star people in their own conscious knowing.* Very tricky, the star people, hm?

Q: (F) P'taah, it has been said the Pleiadeans have shared their technology with some of our scientists. The ones who share the technology, would they be also knowing that we are not quite ready in our consciousness for such advances?

P'taah: But, of course, dear one. But you see, the capability of the people to employ much of this technology is not yet possible. Experiments thus far have, in truth, been very simple. There is much more to come.

Q: The more you are sharing with us, helping us to expand our consciousness, the less I want to know about the technology. To know yes, but..

P'taah: Dear one, the curiosity for the pieces of the puzzle are irresistible and that is alright. It is wonderful and very creative. However, as we say to you each evening that we are together: always,

184

always it comes back to you. If you are not coming into the *knowing,* which has nothing to do with your intellect, nothing to do with the computer residing in your head, but the knowing of your heart, then you are stuck in your own belief structures about who you are, stuck in the non-allowing. [Stuck] in the non-acknowledgment, that every facet of who you are, every thought that you ever had - every action - is divine expression for the experience. It is only your judgment creating agony and pain, it is your judgment, beloved ones, that causes you all to be dying of a broken heart.

Q: (M) P'taah, I would like to know more about the pyramids on Earth. I think there might be a connection between the occult, the pyramids and the people from outer space.

P'taah: But of course.

Q: What can we learn from the pyramids?

P'taah: At this time really nothing, because the technology has gone from the memory of humanity. The way that you ignite the memory is to come into expanded consciousness and all, all knowledge of your planet, your entire history of your planet and of the multiverses will be made manifest to you. You know, this entire thing is about your heart. It is not truly about technology. It is not truly about the pieces of the puzzle which you cannot fit together. We certainly acknowledge the joy and excitement of chasing the illusive pieces of the puzzle - it is grand fun - and there is no judgment about this, beloved. It is a wondrous pursuit, but it does not matter what you do - it only matters how you are *being.*

Q: (F) P'taah, in this lifetime, can we raise our vibrations enough so we do not have to die, to make the translation without having to die?

P'taah: Indeed, it is certainly possible, dear one.

Q: Could you tell me more about this?

P'taah: But we have spoken before about this. *The only reason your body dies that quickly is because you believe that it must.* That is the bottom line. You are so busy protecting your bodies. You are fascinated with how your bodies do no not operate well. You are fascinated with all the germs you may catch. But you are a germ. Indeed, dear ones, you are a macro-molecule of the multiverses. That is who you are. You do not 'catch' anything. *You are not victims.* You really die when you are what is considered infant, compared with other species of humanoids elsewhere, who live normally one thousand of your years. So, if you desire indeed to keep the body functioning, it would behove you well to look at what it is that you believe about your body, about the universe your body inhabits; and it is to understand that the morphogenic resonance of humanity creates this belief structure. So it is to be aware of that and it is called 'flying in the face of the odds'. But it is certainly not beyond the realms of possibility or even probability. There have been those upon your planet who have lived for hundreds of years, as there have been those upon your planet who have survived wondrously without food.

Q: (M) P'taah, you have said the whole point of this experience is to feel the emotions which result from these experiences. Is there any way we can tap into the knowing of what further emotions we need to experience?

P'taah: But, of course, beloved. It is merely to ask. You see, dear one, there is no separation between that which you regard as your consciousness and that which is in truth a greater part of who you are. All that is necessary is to send forth that desire and indeed it shall be. *But it is to allow it to be.*

Q: (M) To what extent have those of us here in this area returned to a Lemurian type presence?

P'taah: Indeed, it is so. The fact that people are drawn forth to this area of your continent and, indeed, those people who are drawn into creating a group energy, as you are doing at this moment, is [due] also [to] the memory of the grand civilization before your own historical times - that which is called Lemuria and the grand mothership of MU[4]. And so this area in its vibrancy, in what you call rainforests,

---

4 According to various sources (one of them being St. Germain) MU was not a land, but a mothership, that is a spacecraft of enormous dimensions. (For example, P'taah stipulates the size of the Pleiadean mothership as being larger than our largest cities.) According to St. Germain, a city and entire civilization was built around MU on the continent of Lemuria.

jungles and its soft and gentle energy and that which is being in the forest beneath the canopy of the flora, is very evocative of how your planet was at the time of Lemuria. A time when the planet was enveloped in a grand cloud covering; when the light was always gentle and the temperature very similar to how it is now in this place and some of your seasons.

Q: So we are actually the same souls who partook of that civilization, returning to a system restoring these energies now?

P'taah: Indeed.

You have a question, lady of lights? Indeed and we will say that this will be the last question, because our woman's body is in stress this evening and so we would not be wanting to tire her.

Q: You spoke before about the flowering that we had on Earth at the time of Lemuria. What happened so that flower withered and died?

P'taah: Dear one, it was an upsurge of masculine energy. But you know, there is no judgment about this, beloved. It is for the experience of it. Everybody gets caught up in what was grand Lemuria and what was grand Atlantis, and the devastation of your planet, and the great drama of the history of humanity. But dear ones, it was for the experience, to see how it felt; and beloved ones, it still is.

There is within the breast of every human the grief about what has been wrought upon your planet, in your pre-history. There has been great guilt and grave judgment and there are many of you who desire to escape into what is called past-life realities and pre-historical realities. But I will tell you this: no thing is as exciting as what you are creating at this moment. So it is not to have judgment. It is to understand that you have created yourself here at this time and this space, for the wondrous excitement to be in every now moment, however it be, beloved ones, and that you will come into the understanding of transmuting your broken heart; that you will come into the wholeness of who you are. *It is called coming home.*

Dear ones, I love you very much. It is honour and joy to be with you always and we would say: Go forth in love and the fullness of every one of your now moments. *(P'taah thanks the hosts.)*

187

Beloved ones, go forth in laughter, it is the greatest aligner of all. Good evening.

Chapter 11

ELEVENTH TRANSMISSION.
Date: 6th of November, 1991.

P'taah: Dear ones, good evening.

Audience: Good evening, P'taah.

(P'taah walks up to Peter Erbe, whose book 'God I Am' has just been released.)

P'taah: Beloved one, congratulations indeed. It is called fruits of the labour brought forth, therein to enlighten, to inflame the hearts of humanity.

Peter: Thank you, P'taah.

(P'taah looks at the man lovingly for a very long moment. Then to the assembled people:)

So, indeed dear ones, well come. And so - at this time in your history - many of you who are stepping forth in your adventure of consciousness become at times overwhelmed with what appears to be greater and greater chaos coming forth in your time. And everywhere you look there is greater devastation, and plans for yet more. Everywhere, when you open up your newspapers, and when watching your television and listening to your radio, there is that which truly makes the heart sink. Everywhere, there is the lusting after power, the greed, and you ask yourself: 'Where will it all end? And is indeed all this talk about the changes to come and the talk of enlightenment merely some remote dream? Or is it something which will happen to somebody else, beyond our time?' Dear Ones, as it is that you are growing in your understanding of how the universe works, how in truth it operates, and as your expansion becomes accelerated - as indeed it is - as your time in your perception is becoming accelerated - we would wish you to understand that, from your new points of knowing, all of what you call corrupt in your environment becomes exposed.

Not that there is more, particularly, but with your own heightened awareness and the growing consciousness of humanity, nothing remains hidden. *Everything is coming into the light* - open for inspection. That which you sometimes think is only a dream, a pipe-dream indeed, may be known by the darkness exposed. As the consciousness expands, as the God/Goddess arises, so is the darkness opened to be embraced into the light. Do not be too harsh in your judgements of those who are the power-brokers. Look at what lies beneath it all. Look at what lies beneath the thirst for war, the lies, the power, corruption, greed, murder and torture. Look beyond what is even greater dis-easement in humanity. It is not there to be judged, beloved ones. It is there to be taken into your heart; *to be embraced, to know that as you may, without judgment, enfold these people into your heart, so indeed, you will change the reality.* Now, know also that everything you behold, that you judge to be dis-easement, to be corruption of power, to be devastation of your planet, all of these things that you judge are part of who you are. A mirror, a grand reflection to you all. So it is not to judge what you perceive to be outside, but to be observed in the knowing of it being the same aspect within yourself. As you embrace that, so you can do no less outside of yourself, and so you are creating the change. And it is indeed to observe in your life how it is that you are shackled to your past; what is called hanging on, what is called superstition, blame, and holding other people responsible for your happiness or unhappiness. Know that every time you blame and every time you judge yourself to be a victim of your past, you are not taking responsibility. Guilt about your past is also not taking responsibility. Responsibility has nothing to do with blame, beloved ones. Guilt, indeed, as we have said before, is merely the lesson not learned. There are those in your world who have suffered guilt for the perpetration of great cruelty upon humanity, but we would tell you that in truth there is no such thing as past. There is only now. When shame and guilt are embraced with responsibility into the *now*, then know that when you are putting forth the thought based on fear and not on love, you feel very discomforted; but when you embrace the guilt and you have learned the lesson [of how it *feels*], you will never feel discomforted again.

That indeed is called the 'State of Grace' - to be in the now moment, in the fullness of the *now*. And so it is with all of the people, and it starts in the breast of each and every one of you. So dear ones, it is not to be in great depression. When you are aware of what is occurring in your world, [then know] that it is the dichotomy of all that your heart desires. We would also say to you not to focus on dire circumstance. As you focus on, and as you are fascinated by [dire circumstances], so you draw to you that to which you give your energy. It is not to hide from, but not to focus upon either. It is to be in the fullness of knowing and the knowing of the 'why' and the 'how'; knowing, that all, *all things* which are not expressions of love are expressions of fear - and even that fear, beloved ones, is called expression of divinity, because *all things* are from the Source, however you may judge it to be. Just as every facet of who you are is indeed an expression of divinity. Now, we will begin with questions, and we would say at this time that there is a microphone and you are required to speak into it, so that every pearl of wisdom that comes forth from your mouth may be recorded for posterity.

Q: (M) Just listening to you I get the understanding that the easiest way to go through those times is just to surrender to whatever comes up. Not to consciously focus on a certain goal that you desire, but to more or less get out of the way and just allow things to happen. Is this so?

P'taah: There, dear one, is a true pearl of wisdom. If each and every one of you could 'get out of the way', get out of the way of what is called reason, what is called intellect, and if at every moment you would be flowing with your hearts desire, with intuition, then of course everything is made very simple. Reason and intellect are not to be judged. We always say, do that which makes the heart sing, follow your heart's desire and listen indeed to your intuition. *(And, jesting:)* Intellect and reason are certainly here for a reason and intellect. That is a joke.

Intellect is a support system to enable you to bring forth the heart's desire. It is called balance, dear ones, you know balance? And certainly, if you can step aside from striving and reasoning, then indeed the whole world will open for you.

Q: So the way to go is the way that feels most harmonious?

P'taah: But of course. How it *feels*.

Q: (M) A curiosity question: It would seem that there is a group of people that runs the world, so to speak, on the financial side and every other way. Are they in touch with the star people and do they work along with them to bring about a faster ending and to carry us on into this change that is coming?

P'taah: Well dear one, that is a very good 'curiosity question', because there has always been - in these last years - very much curiosity about the star people with [regard to] what is called the power brokers. It is so that the star people have indeed been in communication with your scientists and the people of your governments etcetera. Now, we will at this time say that there has been much curiosity about various people from various star systems. There has been some consternation about the ethics involved with some of the things occurring to the humanity of this planet. Now, we will say this: there are many, many civilizations outside this Earth, this planet. Not all of those civilizations are what you would call very advanced. Certainly, technologically they are more advanced than the people here, of course. However, we are speaking of true advancement, and that is the balance of Spirit with technology. Ultimately there is no separation. The grandest technology of all is only able to be utilized by the expanded consciousness of Spirit. Now, this business of coming into the last decade or certainly the last twenty years of this what has been the grand cycle of your time - moving into a higher frequency - we would ask you to bear this in mind: In everything that you are looking at outside of your conscious knowledge at this time, it is very difficult for you to have an overview, because you can only make logical deductions from the very limited boxes of your perception at this time. However, we would say this: humanity has chosen how it is to be. It is up to you whether it be very harmonious or fraught with chaos. Now, needless to say, we understand that you would all prefer it to be a harmonious change.

All things are merely in your own perception, and we have spoken forth before about your perceptions of natural upheavals; that in truth they are of great joy, that they are of grand design, they are cyclic changes - they are natural. And it is so, dear one, that *whereas many of humanity are plotting their own power base in these changes, in truth they are working very well to accomplish what they do not even know of.* It is called 'God moves in mysterious ways' eh? So, indeed, there is much occurring, which people believe are happening for one reason and the overview reveals an entirely different matter and we are asking you to hold firm to what you know in your hearts. Very well.

Q: (M) In 1984 I wrote to the Prime Minister (of Australia), suggesting that since God was legal in paragraph 1 of the Australian Constitution, that his regency of the human soul may be legally inserted as paragraph 1a. The reason for that was then, as it is now, to place an anchor for people to reflect on, that the divinity - or the acceptance - of the human soul was a starting point for the legislature for education. For our children to be educated for living from their soul gifts. For the jurisprudence to be rather for the recovery of humanity than for punishment. My question, specifically, is that if this anchor was suggested again - now - whether the leadership in Canberra would be of sufficiently risen consciousness to accept it now, so that we can cross the bridge between the divine and the material. Would this now work?

P'taah: It is truly not necessary. That which is called consciousness expansion, that which is called the rising of the Christus, will not come from your government. You see beloved, that which is called the Christus of consciousness has nothing to do with what you perceive as law. You see, dear one, the only reason you have law is the fear that somebody will be unlawful. Do you understand? In truth your government is far too busy to worry about the soul, and although God may exist in your charter, the soul does not. The government does not understand that there is no difference, hm? And so, we would say it is not necessary to go to your government. It is only necessary that each and every one of you become so empowered in the light, that you will touch all of humanity and they not even be aware. We do not think that one who is called your Prime Minister is truly interested. His fear of lack of power keeps him very busy. And so it is for all of your governments at this time. But you see, it will be you, each and every one of you, who will be the power-house who will generate what will be true government without written law.

(At this moment the house cat parades in front of P'taah - oblivious to his noble discourse - and he digresses briefly:) She needs no law. You may tell her as often as you wish, she will refuse to hear.

As each and every one of you comes further and further into NOW, into the understanding of your own power, that every moment is a choice of how you perceive anything and that as you come more and more into the fullness of loving that which you are, so indeed it is called 'kiss the government goodbye'. *(Much laughter)* Indeed.

Q: (F) P'taah, in another session you said that we have been all things, like the doer and the victim. I don't quite understand how that could be.

P'taah: Well then, beloved, I will simply explain it to you, because in truth, it is very simple. You have incarnated thousands upon thousands of times. All of you. And you see, when we say there is no separation, indeed we mean truly that, because with each incarnation there is what is called probably eleven twelfths of fragments of soul energy returned to the melting pot, with one thread running through [each incarnation]. So you see, beloved, after thousands of incarnations you are pretty much of a hodgepodge. And so, with each incarnation you come forth for the experience. A different experience each time. Each time to embrace what has not been embraceable before. So indeed, you have been the murderer and the murdered. You have been mother and father; you have been male child, female child. You have been of every race. Hm? Do you understand how it works? You have been everything. *(P'taah turns briefly to a man:)* Indeed, once you were a very beautiful woman, and now you are a beautiful man. This time around in these next years people will speak about population explosion. More and more. Well you see, dear ones, it is a race. Everybody wants to get here to experience the transition. And there are even those who have not been before of human form, who are coming forth now in human form. Indeed. Do you understand?

Q: Yes, thank you.

Q: (M) P'taah, good evening. Those who are coming forth, who have not been in human form before, where will they be coming from?

P'taah: Hm, [from] that which is called whale and dolphin indeed, and that which have been the beings who have been soul energy, light energy, who have not desired before to incarnate in a physical body. There are many, you know, who never felt it necessary. Do you wish to query forth more?

Q: Not on that subject, but I would like to ask you, regarding the dream state, as we spend one third of our life sleeping and we experience the dream state in its various manifestations, could you comment on how important that is or if it is significant in our evolution?

P'taah: Indeed, we have spoken before about this. We would say to you this: It is certainly significant. It is very important. In fact, we would say that in your dream state, all of this seems like a grand illusion. All of this seems like the dream and it does seem unreal. The dream state is multidimensional. Within it, we are speaking of lifetimes, life experiences. We are speaking of travelling forth in consciousness without the body, both on the astral plane and forth to other dimensions, in other galaxies, in other time frames, in alternate realities - that is: 'probable realities' - in what you would call future lives, in what you would term past lives. So you are very busy in your dream state. And outside of this conscious reality you are all functioning on many levels of many dimensions of space and time in probable realities, and you are doing it, I may say, with great grace and ease. But it is also to know that the soul energy is quite impeccable, quite of its own integrity, so you really do not get the 'stations' confused too often. And it is [to know] that nothing can harm you, that you may travel forth in all these dimensions and you may do so at a conscious level, and will not become lost. You will not be severed from the physical body. You are perfectly safe. And you know, dear ones, it is occurring all of the time without you really being aware of it. So we would say you can really trust yourselves, hm?

Q: (F) P'taah, is it quite natural to experience fear at times when our consciousness goes elsewhere during sleep?

P'taah: Dear one, sometimes it occurs like this, when people become suddenly aware that they are somewhere unknown to them. It happens very often when people have a first out-of-body experience in a conscious fashion; that is not in a sleep state; where they are terrified that they will be cut off from the body, but the moment there is a doubt or a fear, it is as if the consciousness plummets back into the body. Many of you here have experienced this.

Dear ones, we will take a break at this time. There are those of you who are on very hard posterior. So it is to stretch to the body, to refresh yourselves and to prepare for query. So we would ask that you be silent for just two minutes. Our thanks.

(After the recess:)

P'taah: And so, dear ones, questions indeed.

Q: (M) Dear P'taah, my question, I feel, is also of interest to the audience in general. Could you, please, tell us about your beginning, your becoming - can you tell us about yourself?

P'taah: Well, dear one, that is called a tall order, indeed.

Q: I know.

P'taah: Do you, I wonder?! Now, you see, that which be I is in truth no different from that which be you, beloved; and that which is this energy that you call P'taah, is but a fragment, as you are. It is also a collective energy. It is not, as it would be called, one person - in a fashion - and yet, there is certainly, and has been for all of your history, a persona known as P'taah. But you see, it is, as we were speaking of a little earlier, exactly the same for you. What you are now, in the persona that you understand to be Self, is merely a fragment of a larger Over-Soul, as it has been called. You are a fragment of the - what is nicely termed - Over-Soul. You are so much grander than what you understand *you* to be, because as we have stated many times: you are all very grand multidimensional beings. You exist in every realm. So when I say to you: 'Who are you?', it is very easy for you to say 'I am..' and give forth a name. For you that is almost all there is. Now, beloved ones, in truth if I am to ask you who you are, you should say: 'That who I be is expression of divinity'. For in this fashion you are expressing more of who you are. As you declare yourself as an expression of divinity, you are allowing that energy to come forth.

Now, when you say forth: 'Indeed I am expression of divinity', so in that moment you are. You know, very soon, when you say it, you will *know* it. We have said to you that it is not of so much advantage to know all that you have been in past lives. It is certainly of great interest, for it satisfies curiosity and that is very valid. Indeed, sometimes it happens, when there is a certain intellectual understanding of facets of who you are, that you connect it to a past life. But you see, beloved one, as you come forth into the light in this time, as you live in the fullness of your now moment, you are also changing every lifetime you regard as past and every lifetime you regard to be of the future. So you see, beloved, when you ask who it is that I be, it is really a quite complex question, yet it is also very simple: Who am I? I AM. I am expression of divinity - and, dear one, *I know that I am.* And soon, beloved humanity, *so shall you.*

Q: (F) Our respects to you, P'taah. I would like you to elaborate on what psychology terms the subconscious, the unconscious, the Ego (the conscious) and the Super-Ego. And perhaps you could explain how you see super-consciousness fitting into it. Is it linked to our physical brain or are actually the neuron charges forming it; and does the DNA system have memory?

P'taah: Now, shall we write a book about these questions, do you think? *(The audience is amused.)* Let us start with what you call unconscious, super-conscious, etcetera, etcetera. Beloved, these are called labels. Indeed. Labels are promoting separation, because you see, dear one, in this grand thirst of intellectual knowing you build boxes. Into these boxes you pile everything, so you may know who and where they are, hm? So it is with *all* of your conscious knowing. It is called 'limited boxes of perception'. The truth is, the more you use these labels, the more you separate Self from SELF. There is no separation, you know? None at all. Now, let us expand a little out of the boxes and labels. The Ego may be likened unto an eye. The eye is an instrument, a tool, in itself it does not know what it sees. It is there to enable you to perceive third dimensional reality. *The Ego was really of the same function, that is, it was to enable expanded consciousness, to perceive what was in third dimensional reality.* In the eons of your history this has changed, so that the Ego now becomes not only the perceptive tool for this dimension, but *it has become 'mad for the power'.* It is terrified, especially at this time, that it will be powerless, useless. In your teachings, you perceive that the Ego will do you grand damage, that you will never make it in the enlightenment stakes, if you are coming from your Ego. Well, you see, *whatever you resist will persist.* The more you fight what is perceived to be your Ego, the more you will find it will get in the way. That which is termed unconscious, that demon wherein lurks a monster out of control, the monster your Ego will keep in place, certainly has taken the place of what is called religion. It has been terrifying, because you have been taught it is not to be trusted. You see, dear ones, you do not trust yourself. None of you, really. But there is no separation.

Your physical body is an expression of the light-body, etheric, soul energy. Without it you do not have a body. Your DNA - DNA/RNA - the great mystery of your times, even that, beloved, is only a physical manifestation. What runs it is called the Source. As humanity comes more and more into lack of separation, *automatically the helices [of the DNA] align. There is nothing in your third dimensional reality which is not powered by the Source.* So you see, your scientists may dissect and photograph and use their instruments of measurement and still they do not find what makes it work. The missing link is called God/Goddess, the All That Is.

With mathematics, astrology, quantum physics etcetera, humanity is really trying to make a map of what already is. Do you understand?

In these coming years there will certainly be a quantum leap within your physics in understanding and knowing. But you see, it will not occur until your scientists will discover their heart. Then all things will be made known.

Q: (M) Good evening, P'taah. My question regards the Over-Self. Is it wise and is it possible to get advise from the Over-Self in the hour of need?

P'taah: Dear one, such is the impeccability of the Over-Soul, the grander Self, if you like - and we are not speaking in hierarchical terms, but in terms of a broader spectrum - that it is always, always giving you wondrous advice that you will not heed. Beloved, of course, you may ask always and *always you will be answered.* You see, there is no separation. That is the understanding you are coming into, that all lies within you, not without. You come and listen to these words, but in truth you know it all. Everything has been spoken forth to humanity since eons of your time. This is not new. All of you must understand, if never [again] one word would be spoken forth, you have it all, dear ones, you have it all already. We will say again: *there is nothing outside of who you are but reflections.*

All the knowledge of all the people who have ever been resides within you. It is not an impossibility that you can tap into who you are. It is real. It is merely to allow. You do not have to do anything. It is there already. *It is merely to be in a very still place and to allow the wisdom to come through.* There is nothing that you do not know. Beloved one, when you are in grief and your life is too difficult to bear, it is truly to know that you are not separate, that you are truly loved, and there is the support of the whole universe for you, *if you will allow it.* You see dear one, everybody, everybody is dying of a broken heart, and all of you are so afraid that it be known you do not perceive that all of you are at the same place.

Q: (M) You mentioned earlier this evening about the big rush of energies to incarnate, to experience something that is coming. Talking to people, there seems to be this time element to get ready for something, and for some reason I cannot feel this. I think we

199

already have what we are chasing. Am I going to miss my train,
because I have this attitude?

P'taah: No, beloved, you will not. You see, you certainly all do
have what you are chasing. All of you are chasing enlightenment and
you already are enlightened beings. But it is true, dear one, that there
are grand changes to come forth, and it is certainly true that your
planet - and all within and upon it - in these next years - are to
experience a grand transition. Now, this is not only on a physical
level, of course. You could say that the physical level is merely an
expression of the etheric. You think only in terms of your own planet.
We are speaking about galaxies beyond the wildest, fevered imaginings
of that which is called science fiction writing. Where do you think
science fiction comes from? *(Chuckles from the audience.)* Wild and
fevered imagination, eh? Is it not wondrous? Do you know that
imagination is what creates your universe? If you can imagine it, dear
ones, IT IS. Now, we must also say that often imagination is given a
little prod. And often it is remembrance. And often it is remembrance
of what you would call future. Very exciting, is it not?

Q: (F) P'taah, throughout our history, have we had what you call
prods? Like perhaps the Spanish Armada being blown away by wind
or something like that? Have we had outside help to keep us on the
right path?

P'taah: Hm. In a fashion, but not quite in *that* fashion. Although
you could certainly say the blowing away of the Spanish Armada was
created by consciousness, but not of extra-terrestrial beings. You see,
everybody always thinks that God is on their side eh? So when you
have a country invading another, the one who wins *obviously* is on
the side of God. Well, that is pretty extraordinary, is it not? You see,
God is on everybody's side. that is the truth of it. When you are
speaking of intervention of the star people into the affairs of your
planet, indeed there has been now and again some intervention. You
know, your planet will never be destroyed. But then you see, dear
ones, your planet is not merely one entity. You will create whatever
you believe, but you know there is never an end to anything, even to
that which you call your extinct species. They are only extinct in your

space and time; that is the humanities and the flora and fauna. They are only extinct in your space and time.

Q: (M) The transition, is this going to be a single event that is to be experienced by all at the same time? Like an explosion of consciousness or how is it going to be experienced?

P'taah: You want me to spoil all your fun? *(Great laughter.)*

Q: How about a little taste?

P'taah: You can have a little taste whenever you want, beloved.

Q: So how is that?

P'taah: It is by experiencing the ONENESS of ALL things. It is called divine ecstasy, It is called transmutation. That is how it will be, and how you create it to be is your adventure. You know we are not a soothsayer, beloved, and when you come to see us next week we shall have a silken scarf draped upon our body with very large golden rings - it is called Gypsy, indeed? And we shall read your palm. *(Laughter.)*

The transition, the coming into the ONENESS of all things, may be as quickly as the blink of the eye. It may be a step forth for all of humanity in that instant, [for all] who desire to experience the change from third to fourth density. Do bear in mind, dear ones, that this is called a label: third to fourth density. For those who do not desire it, they will not experience it. Always, always it is your choice. Next, you will be wanting the date. *(Laughter.)*

Q: (F) P'taah, everywhere we read now in the new awareness books, the revelations and the prophecies, that ten percent of the people of Earth are supposed to survive the catastrophe or transition. What happens to the people who survive this transition? Do they actually stay here, or do they move - as it is said in the prophecies - to some other planet before coming back to help get the Earth on its feet again.

P'taah: Dear me. This is dire circumstance, eh? Now, that which is called prophecy, beloved woman, is not written in stone. You change your reality every moment. Do you desire it to be all people?

Q: I have not really thought whether it should be all people, or whether it should be the ten percent. I just simply want to understand what is happening to those people. Whether those who choose to re-populate or re-furbish the Earth...

P'taah: Dear one, why do you imagine that after the transition the Earth will need to be re-populated? You see, what you are allowing is somebody else's dire vision of what may occur. Now, it is not that there will only be ten percent. It is not that the Earth will be in such a devastated state that it will need to be re-populated. The changes of the Earth herself are wondrous, wondrous changes. That which you consider at this present moment to be dire circumstances brought about by volcano, earthquake and the rising seas, it is only to make room, if you like, for the expanded energy of the transition. It is the Earth moving and preparing herself. In the time of transition there will be many, many more than you can imagine, that is why all the people are coming forth now. Do you think they would come forth just to be annihilated? It is not so, you know. Although there is much, if you like, propaganda put forth and many fear-generating words which are written and spoken about this. I want to tell you, you cannot imagine the wondrous beauty. *The words you have will not describe the ecstatic explosion, nor can you imagine how you will all be in that time, when every atom and molecule upon this planet, and the whole planet herself, will radiate with divine light. Such exquisite beauty is beyond imaginings.* And there will be many, many of you coming forth for this experience. *And when the transition has taken place there will be beings flocking to your planet to sing the wildest hosannas, in joy, in jubilation, in thanks, in blessings.* And those beings will be the unseen ones, they will be the beings from other worlds who do not appear as you do, and yet you will be able to perceive the divinity in all things. It will not matter the shape and the size, you will perceive the Godhood in all. *It will be most magical.* So it is not to be imagining that there will be space ships, flotillas of space ships to remove you, when the 'dire circumstances' come. *You do not need saving, dear ones.* You are beings of grand power, and the reason you are here is to come into an understanding of that power, that you may create the transition as you would desire it, with love and joy and exuberance, in wild creativity, with great honour and integrity.

Q: (M) It sounds like you have witnessed all this before.

P'taah: You see beloved, it has already occurred, and indeed we have witnessed it.

Q: (M) P'taah, do you actually have a physical form as well?

P'taah: Indeed. We are also an expression in physicality. We have been on this planet in physical being, in physical form. We have been with our woman several times.

Q: And is it a form that we could relate to, that we would be familiar with?

P'taah: Indeed. But you know we are very tricky. We can change it. *(Great laughter)*

Q: So you have need for all the things that we have need for?

P'taah: Beloved one, Creatures of divinity, the Gods that you are, *need nothing.* You may desire everything, and indeed you may create it, but you do not *need.*

Q: So you are here for your own learning as well as we are?

P'taah: Indeed, for the grand joy of it; for the experience. Dear one, we would not be here if you were boring us. I know about this. My woman speaks of it very often, being 'bored', hm? You do not bore me.

Q: I was not getting to that. You sound as if you have answers to everything, but obviously there must be areas which you still have to experience yourself.

P'taah: Beloved, but of course. We have said before that many of you think that when you come to enlightenment it is the end of it all. It is only the beginning. It is never ending. The only difference is, you will know how.

Q: (M) P'taah, since we are also engaged in nurturing the feminine aspect of being....

P'taah: Dear one, we wish you all were.

Q: There has always been the presumption that the Holy Spirit is masculine. Surely, to me, it must be of feminine energy. Can you comment on that?

P'taah: Dear one, Spirit has no gender. It is, of course, all things encompassed. It is only your very late religions that have said that God is masculine, and that the Holy Spirit is masculine. It is called balance. It is called encompassing all things. Even you, beloved, are all things. It is just that you do not recognize that you are.

Q: I recognize that I am, but how do I get it all out?

P'taah: It is called allowance. Nothing to do . Merely to allow, and as you love every part of who you are, it is called allowance. Every aspect, every facet and the polarities will be in balance and will blossom, indeed.

Q: (F) Why does there seem to be this block? If we're already enlightened, why can't we see?

P'taah: You have forgotten.

Q: Why?

P'taah: For the experience.

Q: Why does it seem so frustrating?

P'taah: Because you will not allow.

Q: What part of me won't allow? What is that part of me?

P'taah: That which is called masculine energy.

Q: Well, how do I control that? How do I get through it?

P'taah: Dear one, *masculine energy is called control.* We suggest that you read what has already been given forth that you may come to understand this aspect, and then indeed, we would be delighted to speak further with you. Indeed, there will be one more question and then it will be sufficient unto the time.

Q: (M) Would you say that this block that we are all wondering about, that stops us from learning or going ahead or allowing, is a conditioning from when we were very young? Because children, when they come into this world, they seem to know. They seem to be very aware, but we seem to think that because we are older than they, that we are their teachers. We brainwash them as soon as they

are born. This is what I feel my biggest obstacle is - the brain washing I had as a small child. When I was a small child I always wanted to know the truth and the truth I believed was in the church. But it was the reverse for me - it slowed me down. All these conditionings that I had - which are very deep - I am questioning them, and I feel that is allowing me to see things.

P'taah: Indeed and so it is, because as we have said before, there is the knowing, within each incarnation, of the morphogenic resonance of the whole of humanity, and every thing that has gone before. So you build your reality upon your belief structures. The children coming forth at this time have great wisdom, indeed. The veil, as you may say, is not so dense - they shall come forth as great teachers in terms of allowance, in terms of unconditional love, in terms of being able to tap into the knowing within them. However, beloved, we would also suggest to you that you read what has been given forth that you may come into an understanding of how and why it has been created; because we have spoken very much in these last weeks of these questions. As you know, at this time we are preparing a manuscript and so these words are coming forth in semblance of a certain order for an end result. Now, as we have said also before: when you have read and digested what has already been given forth and if you do have more questions, then we are very happy to see you apart from these sessions, so that we may speak fully with you, that you may come to understand what you so earnestly desire to know. So, allow yourself to read the material, so you may come to understand it in the head, although the understanding of the heart is very full. Indeed. It is alright, beloved, you know. Everybody is struggling along trying to understand. *(The man is very moved and P'taah kisses his forehead gently.)* But you know the understanding comes with the heart and not with the head. Still, we would say that we would be very content if you are coming into an understanding of it in your head also. This is called balance, hm? And so dear ones, the time has come, hm? *(To the host:)* Our thanks, dear one.

(To all:) You know it is wondrous joy to be with you, always. *(P'taah pauses, then addresses a certain gent:)* We are very happy that

you are here. *(To all:)* Your energy comes forth to that which is I and it creates a wondrous light. How beautiful you all are. Each moment that you are living in your *now*, each moment that you allow yourselves to be who you really are, you are creating grander and grander light, and all over this wondrous country your light is spreading like the gossamer of a cobweb catching the dewdrop in the morning sun. Soon you will light up your country and that light will be a beacon for the whole world. Dear ones, go forth in light and laughter. Be of light heart. *(To a certain lady:)* It is alright, Lady of Lights, you may tear off your clothing and go dancing in the light of the New Moon. We thoroughly approve, you know?

Dear ones, Good Evening.

Chapter 12

TWELFTH TRANSMISSION.
Date: 13th of November, 1991.

P'taah: Good evening, dear ones.

Audience: Good evening, P'taah.

P'taah: And so, welcome indeed. Tonight is called 'recipe night'. Tonight we are going to speak for a little time about how it is, very simply, that you may manifest all that you desire. It is all so very simple. Truly simple. And you know, in a way, [the fact] that you make it so difficult is only a measure of how you are caught up in the vibrancy of your creation, moment by moment in the reality of third density. It is so exciting, you know, this dimension of reality which you are creating for yourselves. It is vital - vitality - and you know, at the cellular level you are in total understanding of this vibrancy and creativity. Around you, in this area, you are surrounded by lushness, by great creativity of nature. As we have said to you before, it is very clever of you to create yourselves in this place, because outwardly there is manifestation, a mirror for you of how it is every moment of your time - the urgency and integrity of creativity. We have spoken before of how you create your physical bodies moment by moment, without truly understanding that you are doing so; and in this very vital life, you are all caught up in your dramas, in the kaleidoscopic movie that you are making, wherein you are all things. You design every scene. You play the part - of course - you are the star performer. You organize it very wondrously and then you forget it is only a movie, hm? So you become enmeshed in your own creation, most of the time without truly understanding how it is occurring. You do not, as we have said many times, truly understand your power in this creative mode. We have said to you that as you put forth the thought, according to your beliefs, so you create.

Now, this is where we get to the recipe bit: You have within you the ability to tap into the morphogenic resonance of everything in this plane of reality and beyond; of all creatures, of every cellular morphogenic resonance and every morphogenic resonance of the unseen realities - the unseen worlds, the unseen beings. Now, most of the time you are not aware of such. Nevertheless, it is occurring. What you all truly desire to paint on this canvas called your life are beauteous, harmonious pictures and then you wonder why you do not. And so you are wondering about the discordancy that you are creating, not in allowance of the understanding that everything in your life which is not of harmony, is merely there to show you how it may be if you will simply *know,* that within each and every situation is a pearl, a jewel - to be accepted without judgment, in allowance, that the jewel within the lotus may be revealed to you. As you put forth thought, desire, it is to *know absolutely,* that in the putting forth *without the limitation of expectation,* all, *all* possibilities may be allowed to arrange themselves in the best possible outcome for you. Now we have spoken to you, that as you believe that you are not worthy, as you believe that you do not deserve, as you believe within the limited structures of humanity's morphogenic resonance of your culture, your family etcetera, etcetera, so you prevent wild creativity. *As you have expectations of how a thing may occur, you have already limited it.* As you have doubt of the veracity of your own power so you have already limited it. *All possibilities are available to you.* Do you understand? You are so powerful, that as you put forth the thought, unimpeded by doubt, by expectation, by your belief structure - which is non-allowing of creativity - *so the universe will re-arrange itself to accommodate you.* It is truly that simple. It is what is called, dear ones, scientific fact, whether you know it or whether you do not. What we are saying is that it all has a mathematical equation, hm? That is for the technologists. As you desire harmony, as you desire to expand in consciousness, you truly do not have to make what is called ' a big deal' out of it. It is the same with health. In every facet of your life you become fascinated with by the *'nots'.* That is not the tieing of knots, although indeed, you do tie the knots with your *'nots'.* You become so enamoured with the drama in your lives, with discordancy, the excitement and the thrill, that you forget that it need not be like that. You do not need great drama to learn. But you see, for most of you, it is all you know. Tied to the shackles of your past, expectations of people, giving away your power outside of Self. What you are is a symphony, beloved ones, and what you create, indeed, may be wondrous music, because you see, you are every instrument in the grand orchestra. You are the conductor. You are indeed the composer, and you may create wondrous, wondrous symphonies. We would merely suggest that you become fascinated with the beauteous music. As you do so, all that presents itself to you will be of harmony. You, dear ones, are the lyricists, and you may write whatever verse pleases you. You may *change* your reality, but it is to know within you that this is not an intellectual exercise. Your intellect, your ego is there to serve your heart, to serve your intuition. And as you are ruled by intellect, you are again shutting down the possibilities, narrowing down the probabilities. It is not about *doing*. Put forth the thought embraced by the desire. Dear ones, throw it to the universe, knowing in that moment of the thought embraced by the desire, in joy, that it already IS. In that microsecond you have created it. Then there is nothing to do, but to allow.

Now, we are not suggesting for one moment in the 'non doing' that you become what is called a *blob*, that you are to sit down, or lie down and do nothing for the rest of your days. It is indeed to show forth intent, the intent that you *know* that it already IS. It is what you would call physical confirmation. That is where your intellect will serve your knowing, your intuition, your heart. It is to know, that as you have put forth the thought you may indeed *do*, in the knowing that it is already *done*. Now, beloved ones, is this recipe simple enough for you, hm? Many people say: 'I do not know what I want'. That is alright. It is not necessary for you to make forecasts, so to speak, of your life's journey. As you do so you write it in concrete in your mind, and allow for no variation, no creativity. It is alright if you do not know what you [will] want many years from now, or even tomorrow. And do you know, that all around you are the messages of the universe, in your physical reality, to tell you how it may be - what is resonant for you, what is harmonious for you to do in physicality. That is not living by intellect, it is called *living by intuition*. It is called allowing each moment to occur; to resonate to your heart. If you do not know what you want to do tomorrow, that is alright. In the not knowing of tomorrow, you may be in the absolute knowing that everything is absolutely as it should be; that you live in a wondrous and safe universe, that nothing may harm you, that you may create whatever you want. You need not be bogged down by the mechanics of 'how' it should be.

Very well, now. Are there questions? *(The audience remains silent.)* How wondrous. you are all in absolute knowing.

(A lady, a newcomer to these sessions, inquires:)

Q: I am very wary of creating, of visualizing and creating things in my life, in case it is not what is exactly right for me. Perhaps you can tell me if there is a divine purpose in our life. A real purpose. If we start creating when we are not totally attuned, we take ourselves right off the chosen pathway for our lives. It is like a diversion and often a painful one. Is it better just to sit and allow things to happen, and we will just get there in the right way, and probably faster in the end?

P'taah: There is no right way, *for each and every one of you the path is to know who you are, to know indeed that you are an expression of Divinity,* that is all. To know indeed that there is no separation. Dear one, you may visualize. Do you have so little faith in the integrity of you own soul beingness, hm? Of course indeed, there can never be a 'wrong way', there simply IS. And I will tell you this: which ever way you do it, there will be wondrous lessons to be embraced. That is how you grow - that is how you learn. And indeed, it may be as harmonious or as discordant as you choose it. Now, what is discordancy is truly what is not in line with universal energy. When you have aligned and transmuted, the fears that you do not embrace will keep presenting themselves - whatever you do or not do. Once you have aligned and transmuted your day by day fears, never again will they need to be presented, because you will have discovered the pearl of wisdom within. What you call visualizing - a desired event, a material creation in your life - is wondrous, it is great fun. It is called day-dreaming of how it may be. You do it all the time, whether or not it is [about] a 'big deal'. By trying to stop the visualization, the imagination, you are denying the inherent creative abilities within you. And then, beloved, it is also to look at what you believe you do not deserve, because you think you are not worthy. And I will tell you this: You are worthy of everything. There is nothing that you can imagine that you do not deserve, because indeed, who you are is divine expression, every facet of you. When you truly understand this, you will simply fall in love with who you are, and you will know that you could never do anything wrong, because there is no judgment anywhere outside of your head. The universe does not judge, beloved, it simply IS in wondrous allowance of however you desire it to be. The universe does not judge 'good' or 'bad, or 'right' and 'wrong'. It is only you who judge that. You and everybody else. We are speaking of all of humanity. Indeed.

Q: (F) P'taah, when we change our reality to a higher consciousness, will there come a time when we will not want to create on the material level anymore? Will we just want to get above the physical for a while?

P'taah: Dear one, what will occur is that you may express in any realm that you wish. Do not be disdainful of what is called physical reality, beloved. It is a very desired state. With what is called transition, then, because there is no separation, you may integrate what you term to be 'all of that out there' - 'higher realms' (but of course it is not 'higher' realms) - with the knowing of physical reality. It just means that you can play bigger games, beloved.

Q: (F) P'taah, how do you access the knowing of the heart?

P'taah: It is merely to be in allowance. Now, you are very busy 'doing', beloved, but it is not a 'doing' merely to allow. You may indeed put yourself into a meditative state, in that quiet place within, and merely ask for the allowance. That is all you need to do. It is not necessary to go into great rites and rituals. It is a perfectly natural occurrence, accessing the heart. You see, beloved, there is no separation. It is only in the judgments that you have of all of your compartments: that which is your heart, your mind, your subconscious, your superconscious, your spirit, your body, the morphogenic resonance, eh? All in separate compartments. But it is truly not. You are busy scurrying from one department to another, like in your department stores, eh? Shopping around; you have to go here for this and there for that. Well you do not [have to]. You can sit perfectly still and be in all departments at once. Do you understand? And it is in that still place within you, that you may access anything you wish, because it is simply to sit and be open. You may visualize the flower opening, that every portion of the flower is touched by the gentle sunlight, and the sunlight quickens the flower. Every cell becomes alive - enlightened, hm? And you may visualize that every petal is a department and yet the flower is a whole, and that flower indeed is God expressing. The God/Goddess of All. The Source expressing Itself. That is how it is. It is all there for you.

Q: (M) P'taah, I was reading about an aircraft flying in Florida that disappeared off the radar screens. People actually saw it disappear. This was as large passenger aircraft. Ten minutes later it re-appeared. None of the people on board were aware that anything had changed. How is that affected by the 'now moment' that everything is created in?

P'taah: In this occurrence, for a moment it was as if the aircraft came into a time warp. The people [in the aircraft] need not necessarily be aware of the occurrence. And what may seem to be minutes, or sometimes hours, of a disappearance in this fashion, to the people in the experience it is no time. It is merely a time warp. You see, that which is time is very elastic. It is very plastic. In the technology of the beloved craft, which you are all waiting to see so very much, *(the audience is amused, knowing that P'taah refers to the desire of certain members in the audience to see, or travel in, a Pleiadean craft)* there is not only the ability to re-create itself within dimensions, but what seems extraordinary in your technology at this time is, that which appears as a certain dimension of the craft itself in the exterior reality, once the people pass within the craft, they are already entering a *different dimension of time and space. And so the dimensions of the craft are entirely different on the inside, than what is perceived to be the outside.* Those of you who are technologists will understand of what I am speaking. So in fact, it is in this fashion that you may understand how elastic this dimension of reality is, that you set in concrete.

Q: (F) Sometimes, P'taah, when we are waking or going to sleep, time seems to go very slowly, as if I am sitting in this chair and when I go into this space I am not here, I am somewhere else.

P'taah: But it is so.

Q: Is this akin to what you were saying? If I stayed in this state for long enough, would my body possibly disappear and reappear when my consciousness came back here?

P'taah: It is connected, because you see, consciousness in the broader sense travels through space and time. Now, there are occasions when the body itself will translate into another reality. However, we will not be too caught up in this. At this time it is enough for you to play with ideas in consciousness, that you know you may be simply beyond space and time whenever you wish. In fact you are, in your dreamstate, as we have discussed before. You may do it consciously. You may leave your body and travel forth in the conscious state, your body maintaining its own integrity of togetherness, whilst you are gone. You see, beloved ones, that which is your soul energy is not only bound up with your physical body. You are all so much grander than all of that, and most diverse, and most powerful. There is not merely what you think of as your personality Self.

(A lady seems troubled:)

Q: P'taah, what I have to ask now is not very nice. Not very pretty, but I have to find out how you feel about it, because since I saw you last, something has happened that has really made me wonder. If everything is alright as it is, and everything is beautiful as it is, and there is no need for anyone to feel judged or wrong about anything, then how do we explain small children who do not understand these concepts or who probably do better than we do, but cannot explain it, being molested by an adult person. How can all that be beautiful and right?

P'taah: Dear one, this is a very good question, and we have addressed it before. First let me say this: we are not saying everything is beautiful, beloved, we are simply saying that everything simply IS and that there is no judgement in universal law. We are in understanding of the grief which occurs around the children, not only with what is called molestation, but with what is called starvation, that which is called disease. And it is also brought forth with animals, much beloved by humans, when they are maltreated and have no recourse and cannot speak forth of their injury, their anguish hm? And so indeed, what happens is that it creates grand compassion in the breast of all involved. Now, what occurs, beloved, is that in a situation where you are emotionally attached to that which be child or, indeed, beloved pet or any loved one who is victim, one is so caught within the emotion of all of that, that you will simply not understand what is called co-creation. And within the emotional state, the anger, anguish and judgment, one will not take one step back to over view - to know, beloved, that it is not only one life. It is thousands upon thousands of lives.

Now, you know, this is very difficult, because when I speak forth it is very simple to accept intellectually what I am saying. It is not so simple to accept it with the knowing of the heart. Many times we said to you, that you have been every facet of every human who has ever existed - that you have been all: the murderer and the murdered. You have been the child and the child molester. You have been the war monger and the pacifist; and each person involved in an emotionally highly charged event or situation, like that which is child molestation, murder and torture and rape, has brought it forth to come into an understanding of who you are - even the child. Beloved woman, we would ask you to consider this also: You consider the children to be helpless - *they are not!* That which is child is most powerful, indeed. Each situation, especially [concerning] a child who has not structured the beliefs to draw to it - in a conscious fashion - dire circumstance, certainly knows at soul level what is to occur and how. How do you think it is for those beings who create themselves into situations of your planet, where there is no food and where the mother must hold her babe in her arms and watch it die? It is called co-creation, beloved woman. And how is it in a state of war, when the children are creating themselves into a war zone? *(Softly, P'taah adds:)* You see, it is called experience.

Q: Is that why there is so much of it?

P'taah: Indeed. But it also calls forth the belief and there are children who create for themselves an incarnation for the expressed purpose to teach. To teach compassion, to teach non-judgment, to teach embracement - to teach expansion, acceptance and embracement. Do you understand?

Q: I understand intellectually what you are saying. But from what I have been hearing and seeing of it, I think it would be a very difficult thing for human beings - as we are now (no matter how much we believe in reincarnation or whatever), to be able to believe that this is alright; that somehow these things, that happen to helpless beings - animals, children - are not to be judged. Certainly there must be something that we can do about that sort of thing.

P'taah: There is. It is called, beloved: *Love who you are.* It is to know that *everything* that you judge outside of yourself is a reflection. As you accept every facet of who you are, as you align your judgment and as you make it alright to be angry, as you make it alright to cry your tears of rage and anguish, as you make it alright that, in truth, all of you are dying of a broken heart - then indeed you will create the change. You see, dear ones, it does not matter what it is, always, always it is you. Always, always it comes back to you.

Q: (M) Which brings us to the point where nothing is by chance or accident. Could such a child have actually made a soul choice to enter, to experience such a thing for its teaching?

P'taah: But indeed, dear one, this is what we have been saying. There are no accidents, there are no coincidences. None. They do not exist. That is why we said to you many times to allow in the knowing, that all is brought forth at soul level, for embracement, for acceptance - all.

Q: So our duty is to relieve the suffering as best as we can?

P'taah: Indeed, dear one, but it is not called duty. It is to indeed express compassion and to express that which can be of help, of benefit to those about you, hm? It is not duty. If it is done as duty, you may as well not do it. If it is done for the joy of the heart, from your heart, then indeed it is wondrous. It may merely be the sending forth of loving energy to those afflicted, but to know also that the broken hearts are only reflections of your own.

Now we shall take a break. *(P'taah attends once more to the woman, whose heart is troubled:)* Beloved woman, we understand the perplexity and the seeming injustice. But you know, as you perceive the injustice, it is to know that, really, it is only there for embracement and in the larger scheme of things all, all is a path to non-separation, and to God/Goddess, all things. The tears that you all weep, most often they are tears for other people, because you cannot bear your own pain. Dear ones, in the next times together we will indeed speak forth again to you in very great detail about transmutation. Our woman has asked us this evening, and there have been questions within the hearts of many of you about this. So, we will very soon again speak to you about the changement of that which is your agony and your anguish, and the transmutation of that into ecstatic joy and Oneness and non-separation. So, beloved ones, time for refreshment of your bodies, and we shall return very soon.

(After the break:)

P'taah: And so we continue.

Q: (F) P'taah I have a question about those moments when I have to remind myself to come back into the here and now. What I experience then is that I feel my feelings, coming back here. (The lady points to her solar plexus.) During the last weeks I also take it as a reminder to become aware of the Oneness and the feelings change. It doesn't seem to come from here (points again at her Solar plexus), the feelings, but they seem to be around me - often feelings of extreme well-being.

P'taah: Indeed, and so it is. As you bring yourself into the now, into the everpresent, without the past and without the expectations of the future, so automatically you are switching into the feeling of the now. In this fashion you are also opening yourself to experience without the separation. All of that tide of feeling, of emotion that you are surrounding yourself within, can be likened unto a spiral, and so, it resonates on a cellular level and engenders a feeling of Oneness, of wholeness. Well being, indeed.

Q: So they are my feelings too? Just not feeling the knots inside, so much?

P'taah: Dear one, you know that in truth feeling is only energy and it is neutral. It is neither to be judged as a good or bad feeling, It simply IS and it is energy. As you are in the moment of allowance, without judgement of how it be, so the energy centres open that the energy may travel from solar plexus to heart, hm? It may be likened unto transmutation, but it is simply in the being, without the transmuting of the feeling from one section to another. It is only judgment which creates the pain and it is only resistance, only resistance, which is pain. It is just the non-allowance of the movement of energy [feeling]. Pain is not feeling. It is called resistance to feeling.

Q: (M) Thank you for your introduction, P'taah. There is a saying which I have come across in the past called the harmony of the spheres. You mentioned that each one of us is a symphony, and I ask you if you could elaborate on what is the meaning attached to the harmony of the spheres. I was in a situation where I experienced a harmony manifesting as music in such a way I couldn't relate it to any earthly music I had ever heard.

P'taah: Indeed.

Q: So when I attempted to judge it, it disappeared.

P'taah: How extraordinary, beloved.

(Laughter.)

Q: So in a universal sense, could you elaborate on the harmony of the spheres, and does it relate also to the harmonies which exist within the human make-up.

P'taah: Indeed, Now, we have said before to you that what you are is a macro-molecule of the multiverses. Within your physical bodies there is every element found within the Earth, and within your soul energy there is every element of the galaxies. That which is called the harmony of the spheres is indeed celestial energy and it certainly may be heard as music. It may also be perceived as colour, because that

217

which you understand to be sound and colour is of the same light frequency. Now, when you tap into a state of beingness you may do this consciously, or you may indeed inadvertently open to it. It is the feeling that engenders the symphony of the spheres, but as you are the macro- microcosm - macro-molecule - so indeed within you is the knowing and the reality of that harmony. It is wondrous, and you may create the harmony of the spheres within you, beloved, in an inadvertent manner, and we would venture to suggest that you will not judge it again. Imagine what you may do, when you come into the knowing of who you really are. It is also for many of you, that you will smell sweet perfume. That is also a spectrum of the harmony, and for some of you , you will perceive what it is to be every facet of the grandest crystal that you could imagine, with every colour spectrum portrayed wondrously. You are very wondrously powerful - all - and you know, when we speak of these things, it is no fairy story. It is truly reality. We understand that when you become despondent, you think it merely some placebo, some straw in the wind to grasp onto, that you may bear your days. But we will tell you, dear ones, truly it is magical world.

Q: (M) P'taah, I would like to pay special tribute to the spirit of the Aboriginal people, particularly to the jungle and inter-communities in the Kuranda area. For those who have become overwhelmed by some negative situations, how may we best harmonize with the Aboriginal people.?

P'taah: Dear one, it is exactly the same as before, when you asked about the government.

Q: Are you saying it is not necessary to...

P'taah: You do not have to DO anything. It is simply to BE, to know who you are. To harmonize with you, to understand that there is no separation, to learn how to love who *you* are. That which is called Aboriginal, in what is perceived to be the trials and tribulations of this race, it is a grand lesson for both Aboriginal and for all who are perceiving what is occurring. And we would say this, beloved: in spite of what appear to be insurmountable odds, that which is called Aboriginal is coming into a grand understanding of who they are. We are not speaking so much of a conscious level. And it is very much that Europeans - or the new-comers to this country, non-indigenous people of the continent - in their concern about and interest in the ancient race and the knowledge all but lost to conscious memory, are igniting not only the soul memory of the Aboriginals, but of all of the people who have been Aboriginal. On the continent of the Americas it is the same situation, where a culture, a race of people who have been very grand in their knowing, have lost their culture, but have gained something else. There are the people coming forth at this time who have known lifetimes of people all over your planet, in the knowing of the soul energy of what it has been. It is the same on the continent of Africa. Dear ones, in truth nothing is lost, it is not that you have to scurry about to save anybody. But indeed, it is to be in honour and respect of all people. But you see, it is very difficult when you are not in honour and respect of who *you* are. So again, you see we come back to you..indeed.

Q: Thank you. One more question regarding the pain in cancer patients: If their pain is resistance to feeling, how may we best relieve that pain?

P'taah: Again, dear one, what it is that you are giving forth is energy of healing, but truly the only person who one may heal is oneself. And indeed, it is first to know that each creates their own reality and until such time as people will take responsibility for their own creation, it is very difficult for them to align what they have created; to align the judgments which have created the dis-easements in the first place. Do you understand?

Q: Yes.

P'taah: So indeed it is worthwhile to make known [to the patient] that everybody is responsible. That nobody is a 'victim'. You do not catch a disease, hm? It is not a stray dog. And [to make known] that you create any physical dis-easement merely as the mirror of what is occurring emotionally. *Responsibility relieves you all of being victims,* and in this fashion, it may be done gently, for that which we are speaking of is not to be used as a sledgehammer. And if indeed you are giving forth words which are not received, it is alright, because dear ones, it is not for any of you to judge how another may create their reality. As we have said, from the compassion of ones heart you may send forth that loving and healing energy, but the bottom line is, if someone does not wish to be cured, then they will not. If somebody has decided it is time to translate, through whatever belief structure, then it is their privilege to do so. They are writing their own movie, their own symphony, hm?

Q: (F) Last week you spoke about species of animals and plants that are becoming extinct in the world, and you said they are not disappearing, can you explain this?

P'taah: Hm, but of course, this will relieve your minds. Nothing is ever lost, and what appears to be extinct is only extinct in your space/time framework. *Nothing* is created merely to disappear into nothingness and this again is talking about over-view of how it be in your worlds. There is not merely one Earth, beloved, as we have said before, the reality is not concrete. It is malleable. there are many Earths, many, many, many realities - more realities than there are people on the planet, and that which appears to be extinct is only in this space/time framework. Also there are many species of creatures that you believe to be gone forever, who in fact are very happily established not only on other Earths, but on other planets as well. There are many wondrous, what you would call, conservation offices. You are not the only ones who care about what is occurring with your flora and fauna. There are many species of creatures who were no longer necessary in what may be termed the evolutionary patterning, but who have gone on to create a different reality for themselves. Frequently, when you ask forth your questions, it is multidimensional in so far as you have noticed, very often in your readings of different entities and energies, that sometimes answers seem very disparate. Well, often it is because there are many answers. When we speak to you, we are speaking forth in what we perceive to be clear and simplistic terms within the reference of your own boxes of belief structures. But we are also attempting that you would understand that for every question there are multitudinous answers and we do not wish to clutter you. There is enough already given forth which is confusing for humanity. You get enmeshed in what is called the *doing*. You become fascinated by the ritual of *becoming*, and we would wish you to understand that what you are *striving to become, you already are*. And you are indeed the central sun of your universe and whatever you perceive to be outside of yourself is simply a reflection, a mirror. Ultimately, whatever you do, you will bring forth situations - whatever storyline you chose - for you to align, that you may become whole - that you may come home.

Q: (F) I would like to ask you, would you agree that our mission here on Earth is to transmute Spirit into matter; and if so, what is a nice, quick, efficient way to do it?

P'taah: Dear one, you already are Spirit incarnated as matter. What do you think you are? You are grand spiritual beings, who have chosen to incarnate, who have chosen to experience the third density of reality. You already are grand spiritual beings. There is nothing to do. There is no quick way to get to where you already are. You see, you forget. That is all. But you have a vision of being that poor and lonely personality Self, struggling to be spiritual. Well, now you know. You are in truth a grand spiritual being and you have chosen this for your own experience and when we speak of transmutation, beloved, we are not speaking of transmuting into what you already are - that is Spirit into matter. We are speaking of merely transmuting agony into ecstasy, separation into non-separation; and that is very quick and easy - it happens in the blink of an eye. You merely take responsibility for what you have created; you align the judgment you have about yourself, of the situation and everybody involved in the painful situation. Know that pain is accumulative and is only resistance brought about by judgment - allow it to be and know that whatever is, is valid. Simply because it is. And presto - the energy centres open and you experience Oneness. But you cannot DO it. That is the grand dichotomy. You cannot do it, you may only allow [it].

Q: But how do you allow it, instead of doing it?

P'taah: Know that it simply is, and know that whatever is, is an expression of divinity.

Q: (M) *So we are already there, and also, if we rush out to help someone who we feel is suffering, we are actually robbing him of the experience?*

P'taah: In a manner of speaking, we could say in extreme cases, indeed. But dear one, we are not saying do not help and do not follow the dictates of your heart. We are not saying at all that you should suppress compassion and the desire to help a fellow man. We are merely saying: Recognize and acknowledge that everybody is creating their own scenario. You see, as you are within the frequency of whatever is occurring, so you are there to learn your own lessons. We are not saying to disregard people and not to give forth help and compassion and healing and wondrous energy. That is what it is all about. It is really not to be in judgment, when you are giving forth help and [not] to have expectations of how it should be. That is all. Do you understand? But indeed, we say this also: If you are trying to change somebody else's experience, you are merely creating an opportunity for your own learning.

(Another gentleman, a regular participant in these events, is about to ask a question, when P'taah walks up to him. He stands very close and their eyes meet for a long and very loving moment. P'taah breaks the spell by teasing:) Do you want to kiss me?

(The man, mischievously:)

Q: No.

(Chuckles in the audience.)

P'taah: I am very disappointed. *(And, referring to humanity's common image of extra-terrestrials:)* Is it because my scales are turning you off?

Q: No, it is not.

P'taah: We are very happy about that.

Q: But I love you anyway.

P'taah: But of course, beloved.

Q: Yes. Anyway, the question I wanted to ask..

P'taah: Never mind the question. Let me tell you how much I love you too. And I do!

Q: That is beautiful.

P'taah: So are you.

Q: So let's forget about the question.

(The audience is aware of the subtle, delightful game the two are playing with each other and there is a lot of laughter.)

P'taah: You know, you know the answer already, however, beloved, ask your question.

Q: Just for the joy of it.

P'taah: Indeed for the joy of it, always for the joy of it.

Q: That is the only reason I ask questions anyway.

P'taah: I know.

Q: So, okay - the question: If we are co-creating situations with whatever partner is involved, what about Karma, if everybody is an equal partner in it?

P'taah: Hm, what is Karma?

Q: Well, I understand Karma this way: If you do something wrong, you have to correct it in another opportunity.

P'taah: Indeed. Well, how do you think it is - this Karma - which is called punishment for wrong doing, if there is no judgment in the universe, beloved?

Q: That is what I was wondering about.

P'taah: Then you may answer.

Q: Then there is no Karma.

P'taah: You see, perfect answer. Top of class. That which is called Karma is an idea construct from old religions. And you know well about how idea constructs work. Now, an idea construct of one of your newer religions, which is Christianity, says that if you do wrong you do not have to wait for the next incarnation - for indeed, they do not believe in it - you simply go to hell. Quick and easy, eh? A one-shot affair - if you do not behave you go to hell. Well, I will tell you this: For those of humanity who believe it - when they translate from this physical reality - that is exactly what they will experience: hell. In whatever way they have imagined it. Then they will be bored and come into the knowing that they can change it - any moment they wish.

Karma - *in this fashion* [as punishment] - does not exist. How can it possibly, when all of the incarnations are in fact occurring at the same time. That 'knocks that one on the head', does it not? You see, it is technically impossible, because outside of this time frame it occurs all at the same time. It is merely this: whatever experiences that you do not embrace - without judgment - in any one incarnation, you will in another; but that is not punishment, because there is no right or wrong. You are simply either in harmony or without harmony. That is, whatever is, is within the universal law called love or what is called fear. So it is simply an experience without any judgment. And we will say again to you: You have the opportunity to change not only the past and the future of this lifetime, but also of every lifetime, because there is no separation. As you experience the transmutation, that is exactly what will occur for you.

Q: (F) P'taah, back to the flora and fauna of this place: There has been a report that a whole lot of whales have beached themselves again on one of our beaches. Why do they do this?

P'taah: How many answers do you want beloved? There is physical reality, and there is - more importantly for you to understand and I will say it again - that which is Cetacean, that which is whale and dolphin is of the same soul energy as humanity. Whales are wondrous multidimensional beings, the historians of your planet, who indeed hear and give forth the harmony of the spheres. They indeed communicate with the star-people and have always done so. They are certainly expressing to humanity wondrous opportunities to learn. And as it was before, when there were those of Cetacean who were stranding themselves within the arctic pole, the nations came together to work in harmony. One of those whales, indeed, ascended and gave forth light for the whole world. Each and everyone of you may join with the consciousness of Cetacean. We have said to you, how you may do this with any species of flora or fauna and even that which you consider to be inanimate objects. It is merely to go to the quiet place within to ask forth that you may be joined with the consciousness of whale or dolphin. That which is whale is the over-soul of the dolphin and you may join consciousness, and in this fashion the wondrousness of the soul energy of Cetacean will speak forth to you. Thus you will come into a wondrous understanding of unity, because Cetacean indeed knows what is joy, knows what is unity and sharing and wonderment at the planet. *They know themselves as absolute expressions of the Source.* To know that there is no separation; there is truly no separation between you and the God/Goddess or any species - whether here upon the planet at this time or perceived to be gone. On any planet, in the seen or unseen universes. It is all there for you. All that is necessary is that you ask and allow.

Dear ones, all is within you. I am only here to remind you of what you already know. It is certainly joy for me to be with you in this fashion. We would also remind you: The greatest teacher you have is you. Everything, everything that you desire is within you. Yours is the power and the glory.

Our thanks, dear ones. (And tenderly:) How beautiful you are. I love you.

Good evening.

Chapter 13

THIRTEENTH TRANSMISSION.
Date: 20th of November, 1991.

(As usual, P'taah takes his time and looks at each one in the audience.)

P'taah: Good evening.

Audience: Good evening, P'taah.

P'taah: You know, how you create your reality moment by moment is certainly not [done] conscious[ly]. Your over-soul, the greater of who you are, knows indeed what you are about. In this fashion you may cast it all into the universe and know that you are creating it for the greatest joy of who you are. There is nothing to do. It is simply to BE in the moment.

(P'taah addresses a lady-friend of Jani King, knowing that on this particular day the energies in Jani's house gravitated around the house-pets.) So, dear one, how are the new babes of the house?

Q: Wonderful now, thank you. There was a slight drama..

P'taah: Indeed, but you are a queen of drama.

Q: Absolutely.

P'taah: So, of course, each and every one of you creates in your day-to-day life - in this vibrant third density reality - that which gives excitement. It is merely that you do not understand truly that the excitement may be one of great joy and not of that which makes the heart heavy. It is that simple.

(P'taah now turns his attention to another lady, who is about to embark on a three month journey to India.) And so, beloved one, wondrous journeys to come forth, hm? And casting off trepidation and fear of the future, to know indeed that you are creating wondrous adventures in your journeying of the physical body. No less than the adventure of the soul energy, adventures in consciousness. And you

may go forth into other cultures and indeed, beloved, it is to know that you go forth for an adventure and what you consider to be learning; but in truth you are to be imparting great knowledge as you journey forth. So you will, indeed, have a wondrous time - be not concerned with the outcome.

And so, each of you is on an adventure. Each and everyone of you is calling forth the greatest learning and as these days and months and years quicken into the changement of cycle in your reality, so you may choose. At each moment you may give forth the essence of the divine expression you are. Call forth - in these times of change - adventures in joy, in laughter, in love, instead of buying into what is predominant upon your planet, which is fear and devastation. Now, beloved ones, indeed in this time to come you are writing forth a culmination of fifty thousand years of your history. You know, we said before that you have created yourselves into this incarnation that you may bear witness and that you may contribute to the changes which are coming, the changes to occur on your planet. You do not have so much time to wait - *it is imminent.* For each one of you there is a yearning within your breast to experience these changes, to experience the change in the consciousness of humanity; the *change* within your Earth itself. And she indeed is quickened. The coming changes are merely dewdrops on a blade of grass. The Goddess that you call your Earth knows indeed what is to occur. And so do you. *So do you indeed.*

Now, we say this to you: For many of you, there is frustration, because you are not at an intellectual and conscious level of understanding to know [consciously] what is occurring. There is the grand frustration, which says: 'We know that it is all out there, but we cannot fit together the pieces of the puzzle.' Dear ones, it is not necessary, in fact, truly, it is not even possible. We said to you before that your intellect and your personality Self, the Ego, is not able to encompass the essence of the God/Goddess of the All-There-Is. *The only part of you which is able to encompass all that is to occur is your heart.* Your heart taps, absolutely, into divine essence, the knowledge of all time continuums, of all galaxies, of everything that in your perception has already occurred and all that is to occur. But it is your heart and not your intellect. It is not your conscious understanding. Now, in a way, this is called a leap of faith; it is to be in touch with the feminine polarity of your being to allow intuition, to allow the allowing. It is not possible for you with your intellectual mind to encompass all that is to occur. The act of faith is to *know* that you may *create it as you will*. This is called your own sovereignty. It is not about *doing* anything. It is only about *allowing everything*.

(For the reader, who did not have the fortune of personally participating in these events, the following comment seems appropriate: P'taah's communication, from here on to the point when he calls for a break, is permeated by an unusually elating vibration, giving attributes such as softness and tenderness back their original meaning, stirring a deeply joyous remembrance; subtle, yet powerful qualities reaching directly into the hearts of the audience. It is difficult, if not impossible, to describe the waves of love carrying his words. It is as if the ones assembled form one soul and that soul is laid bare, only to avail itself to his ever so gentle, soft and healing touch. It is Being touching Being directly. It is true communication. It is an experience.)

Now, we may paint a picture for you about how it is to be, the glory of it all. But you see, when we speak the words to you, we are only speaking in one dimension of reality. Your soul energy - your heart - understands on *every* dimension of reality. And so we could paint the picture, but it would only represent an absolute microcosm of what is called infinity. In this life, beloved ones, in this time, there is always choice. We talked before about your conscious perception of how it [may] be, that as you are in lack of love of SELF, as you have a sense of unworthiness, so you limit all of the things you may be, so you create separation one from another, and each from the Source. The choice at each moment is between fear of not being enough, fear of being unlovable and unloved, fear of being unworthy and the fear of every translation [death] - and [between] love. Always, always this is the choice. Everything in your life, every ill, every dis-easement only reflects your choice of fear, your choice of judgment. When you choose love, when you choose non-judgment, know that everything is exactly as it IS and in the Isness it is called divine expression. So you are putting forth into your universe and the multiverses that you have chosen love, [to be] in allowance, and every time you do so, you are painting a brush-stroke on the canvas of your reality. Each time you choose love over fear, each time you choose love over your desperate desire to DO, you are changing this reality.

We spoke to you about transmutation and for many of you still it remains a mystery. We will speak of it once again. Transmutation, the changement of every agony, anguish and pain that you have felt within your breast, the anguish of ten thousand lifetimes - the dying of all of you of a broken heart - to change it into divine ecstasy, into unity, into Oneness. *Know that you are not - and never have been - separate from your Source.* You truly have never been separate from each other. You have never really been separate from all of the brothers and sisters you have on countless planets throughout your galaxies. You have never been separate from any creature of your planet. Not separate from your Sun and your Moon; not separate from any leaf, or blade of grass, or bloom that flowers in your garden. You have forgotten, that is all, and in your pain and in your judgment of who you are, you have closed down. You have forgotten that who you are indeed is God/Goddess smelling the rose of the vibrancy and excitement of this dimension of reality. *It is your judgment of this reality, your judgment of who you are that creates the pain of resistance within your breast, that you forget the unity, the Oneness,* that is all. Beloved ones, Universal Truth is called love. It is so simple, really. It is called LOVE, and that love IS the loving of who you are. Know that however you are, whatever your thoughts, whatever your action, it is called expression of Divinity. The Divine Source is creating everything seen and unseen in all the galaxies, in all the multiverses and you see, you think that you are so insignificant, you think that you are so bad, so unlovable, and in that invalidation, lifetime after lifetime, from your childhood on, you have become invulnerable. And that is alright, dear ones. It is alright. *The only thing that is between you and the Divine Ecstasy is judgment.* It is judgment which creates the pain, the anguish, the agony that you have all known lifetime after lifetime. The pain, beloved ones, is not a feeling. *It is merely resistance [to feeling], because you have become too afraid to feel.* What you are, Divine Sparks that you are, indeed created from imagination and emotion, that is who you are. You have created yourselves in matter from consciousness, from the integrity of your soul being, to be united again with the Source. Beloved ones, do not judge who you are. *Who you are is wondrous. Who you are is beauty beyond measure. Who you are is the Source expressing in this density of reality, and you have chosen it.* It is to allow it all. To know that each time you bring forth a situation that creates pain, you have created it that you may understand unity. You have created the pain, the anguish with your loved ones, [the pain] with what appears to be circumstance, you have created every discordant situation that brings you pain, that you may say: 'Ah'. Within the pain is a lotus flower that only requires non-judgment and the warmth of love for who you are, to flower and reveal within the emerald of your heart, in unity, with Divinity. It is you who have created it all. There is nobody out there doing it to you. From the grandeur of your being you will continue to bring forth situations that you may embrace, to open [yourself] to who you are without judgment. It simply IS, and in that ISNESS - if allowed - you create the change. That is called transmutation, to change agony to ecstasy. Beloved ones, *it is the miracle of humanity. It changes everything, every cell in your body.* It changes the past and future of this life, and every lifetime that, in your perception, you have already lived and all the lifetimes in your perception that you believe to be your future. And this is done merely by allowing. In the allowing you are creating the Golden Age of humanity, coming forth after fifty thousand years. It is your choice moment by moment: love or fear. And dear ones, it is not to push away the fear. It is not to try to 'get rid of'. In truth it is really not even to try to change, it is to know that even the fear, even the judgment you have about yourselves is also valid; it is also divine expression, and the only way you create the change is to gather it into the light of you; to know that it is alright. It is likened unto holding the child that you really are within your arms. To caress the child that you are, to say 'it is alright, there is no thing to fear'. To embrace it all, all of the judgments you have about yourself, all the judgments you have about other people, that are merely in truth reflections of how you are within yourself, [how you are feeling] about yourself. In this fashion, you align every judgment by the allowance. You create the change by the allowing, not by the desperate striving for change, the invalidation yet again of who you are. It is by knowing that every step you have ever taken, every reaction in fear and terror, everything that you have judged to be dreadful, petty and mean, that have been reactions of jealousy and possessiveness, every reaction you have taken against somebody else in truth has been against you; has been an invalidation of who you are. Beloved ones, it is like a child learning to walk. You do not strike the child because it is not yet knowing how to put one foot sturdily in front of the other. You do not beat the child. You embrace the child and you say 'well done, beloved one, it is yet another step you have taken'. And so it is that you may be with who you are, and in the embracement and allowance you open up the energy centres within you and that which has been the pain is allowed to move from your belly to your heart, from your heart to your crown, to create the fireworks that reverberate through the galaxies. It is called I AM. It is called God/Goddess. It is called unity and non-separation. It is called I AM God/Goddess, All-That-Is, expressing in this wondrous and exciting, vibrant realm of reality. And you have created it all. You are wondrous beings.

Transmutation, beloved ones, is your miracle. Take responsibility. Bless yourself for every creation of dis-easement, every creation of agony. Bless it, because in the non-judgment, in the acceptance, you are discovering indeed the GOD you are. In the allowance of it all, merely to be without judgment, without trying to change anything, you are creating the rainbows and the starlight. So be it. *(There is a long moment of stillness.)*

Very well, we shall take a break that you may contemplate that which is within your heart. We shall return very soon to you.

(After the break:)
And so, dear ones, Time for questions, eh?

Q: (M) Recently there was a rising of women who aspired to the ministry, the priesthood in theologically based organisations. Basically, they were refused. My feeling would be to support those women, yet is it necessary in the overview to be refused, so that those barriers of those religions separating groups of people come down? Or was it in fact a true rising of feminine influence to create balance in those religions, or a cross-over into a masculine effort? Could you comment on that?

P'taah: Indeed, dear one. Now, of course, that which is called female in gender has nothing to do with masculine/feminine energy within each and every individual. The making of laws to establish equality is really only a reflection of the rising of feminine energy. Female gender wishing to - what is seen to be - encroach on what is masculine territory is something else again, eh? We have said before that what is called 'women's movement' is really 'overkill', because it is still masculine energy. We would also say that your religion, whatever name it has, whatever sect it may be, is still called a reflection of your history. *Religion has enslaved and enchained humanity for thousands of your years, creating separation, creating invalidation, creating judgment of that which be right and wrong, creating fear in that you are so sinful, that you are paying for your 'original sin'.*

Now is the time of changement of all of that. It is time for humanity to understand that there is no judgment in the universe. There is no judgment outside of your head. There is no god, or what is called hierarchy of angels, standing on a cloud judging humanity. There is no right or wrong in the universe. There is no good and bad in the universe. There simply IS, and whatever IS, is valid. It is all divine expression. It is all called opportunity to be embraced into love. So anybody striving for whatever, within the framework of archaic institutions, is really trying to close the stable door after the camel has fled forth into the desert. Dear ones, your future is not tied to institutions. In a way you may say there is no future written, but in the moment, each moment of your heart, you create in a framework of love and unity your next now moments. It is to go forth and do that which makes the heart sing. It is called spontaneity, it is called the wonderment and joy of what you may create. *When you tie yourself into institutions of religion, you are enslaving yourselves to the shackles of the past. You are denying your own free dominion, your own sovereignty.* And if you decide that you wish to do so, well indeed, it is valid. Indeed, it is always your choice. But it is not necessary.

Q: (M) Good evening, P'taah. I have trouble with the concept that there is no right or wrong. Do you mean in the long run, or in eternity, or in the eternal now, or do you mean in our realm of existence?

P'taah: Dear one, there is only the eternal now.

Q: I mean, if people murder other people for gain, in my personal opinion it wouldn't be good, because there is no love in it.

P'taah: Indeed, but you see, dear one, humanity has always chosen that which be called love or fear. If one would take the life of another for gain it is the fear of being without, and that is what we come to: judgment. It is to know, indeed, that whatever is not an expression of love is an expression of fear, and it is valid. It is also divine expression, beloved, because there is nothing in the multiverses that is not created from the Source. It is indeed to take one step back, to be in understanding and not in judgment. To know also that what you see outside yourself is a reflection of who you are. And each and every one of you have been everything, every expression. There is murder within the hearts of all of you now and again. That is alright. It is to embrace the fear which creates the thought. Why do you think it is that you have all of these laws? You have them because you are terrified of who you are, that if you are to let run amok, there will be nothing left. But you see, dear ones, within all of you there is also the spark of Divinity, the Christus, the Source. It is the light filament (DNA) which connects you with everything and everyone. And you close it down in fear. Dear ones, it is not just fear in this lifetime, it is accumulative, fear and pain. It is programmed into you. But now you may know that you may change it all. And now is the time. *It is the ripeness of time, not only within the consciousness of humanity, but in the living consciousness of every atom and molecule, of that which you regard to be inanimate, that which be flora and fauna, that which is the goddess, your planet herself.* The change resonates throughout the multiverses. It is not only this planet at this time, because as we have said to you before, outside of this space/time continuum it is all occurring at the same time. In all your thousands of expressions of your planet, in the thousands and thousands of expressions of humanity, all, ALL are quickening for the change, for this wondrous future that YOU are writing every moment of your being, in the ISNESS of now, to come into the allowing of the God/Goddess I AM.

Q: (M) Greetings, P'taah.

P'taah: Greetings, indeed, beloved.

Q: I have a real problem with - ah - remembering my question. (much hilarity.)

P'taah: But you see dear one, it is only to show you that in truth the heart of you has no questions.

Q: Yes.

P'taah: And we would say this also, for all of you, that every word that I say forth to you has been said before. Everything you know. And it is merely that the words spoken forth ignite the knowing within you. You already know it all. I am truly quite redundant. Hm.

Q: That seems to be the same thing. While I was thinking of questions to ask I already answered my own questions. But still, the question is open: How can one calm down the thinking, the mind, you know? I like thinking of good things, of pleasant things, joy, but all the time there creeps in the little voice in there that says some funny things. How can I control that? You know what I mean? Of course, you know what I mean.

P'taah: Well, the first thing is to know that you truly do not have to control anything, because control is masculine energy. Hm? To go forth and conquer, eh? You hold down those rebellious thoughts, push down what is called Ego, subdue that which is called intellect and put your mind to sleep forever. Well, that is all very well, but you see, it is not balance. It is called 'masculine, to DO'. Now, you may allow those rebellious thoughts, hm? You may allow your busy mind. It is simply to ALLOW. That is called feminine energy, receptive, allowing, nurturing. It is in the allowing that you will create the stillness, that you will create the void, in which to be. Hm? You know it is likened unto a spiral, the energy of all of you. It is that every molecule is a spiral of energy connecting you with the multiverses, and even the words that we speak forth to you, you may envision as dancing on the edge of the spiral. Now, those which are the words, the ideas, the intellectual understandings, all in the limited boxes of comprehension, are truly like light filaments connecting with each other on the perimeter of the spiral and within is the void of creation. the void of allowance, the void of total infinity. You do not come into BEING by trying to control. *Allow.*

Q: I understand. Are we in fact our thoughts. Is it so?

P'taah: But beloved, you are also so much more. You are so much more than your comprehension. When you say thought you are thinking of intellect. You are thinking of belief structures. You are thinking of the personality Self, but you see beloved, you are so much grander than all of that.

Q: So every thought we have is okay?

P'taah: But of course, dear one. *Whatever it is, is alright.* Merely allow it and know that within is what is called the integrity of soul energy. It is the fervent desire of the heart to come into allowance, to come into unity, to Oneness, and know that every time you are trying to control, you are shutting down spontaneous creativity. Alright?

Q: Thank you.

Q: (F) I would just like to ask something to do with the fear bit. It is a personal thing, I suppose, because it has to do with disease. For me, I realize it is fear and lack of Self love. I go along thinking I am allowing and that I am doing alright; then I find that something has manifested in my body. So I think it is obviously there to show me something more and then I look to see what I am missing. I mean if I just sit back and allow, I feel: 'Well, I was doing that before, and then this happens'. So I start looking to see where the fear is, what

is in the subconscious mind that is creating it? That is the part that I get confused about.

P'taah: Indeed, and it is also to understand, beloved, that the humanity of this day has within it fear of the subconscious. It is the fear of the monster that you have no knowledge of, that lurks beneath consciousness. Once upon a time, not too long ago in your terms, humanity was able to give all of that fear to religion, 'faith in God'. No matter what the chains of religion were, there was always somebody to give it all to. In the changes of these years, people are realizing that there is not some old man up there called God, who is very busy judging who you are. Then there is psychology, and you are taught there that within your subconscious lurks a raging monster who could do anything. That is the other side of the coin. Your subconscious is not an enemy, there is not a monster creating things in your life that you have no control over. It does not work like that. It is indeed so, that diseasement within the body is merely a reflection of emotional dis-ease.

Q: Well this is the part. If that is it and I recognize that, then how do I know that I am dealing with it?

P'taah: It is to have faith in who you are, in the knowing that as you wish the change, it is to embrace diseasement.

Q: Just recognizing it is enough?

P'taah: Indeed, but it is to love - without judgment - the diseasement, because within the disease is the jewel. Without judgment. To know indeed that you have created it, that you may change it all, but you will not change it from fear of the diseasement. You will not change it if you are in judgment of who you are. Do you understand?

Q: Yes, I do, but sometimes it is so hard to know. So then I have to look at the areas in my life, where I am not paying attention, I mean, I have to look at myself to a certain degree, whether it's in my diet, or whatever.

P'taah: Dear one, it does not matter what you are eating or drinking. It only matters how you feel. *Your reality is created from feeling.* Emotion creates the feeling. Hm?

Q: Right, well just recently I have been feeling really exhausted all the time, and here I am thinking I am doing the right thing, and I am tired all the time. So somewhere along the line something is not working. Is it that I am lacking in vitamins, or am I not dealing properly with something?

P'taah: Dear one, if you are lacking in vitamins it is indeed because you are not doing something properly, and it has got nothing to do with what you put in your mouth.

Q: Well this is the point, to find out what I am not doing properly.

P'taah: Beloved, you are not living in joy.

Q: (Laughing) Well, I'm doing my best.

P'taah: And the more you laugh the more joy you are in, and the more wondrously vibrant and healthy your body becomes.

Q: Yes, I guess it is just going along.

P'taah: Indeed it is just going along. Beloved ones, we are in understanding, absolutely, how it appears to you, how it appears to be difficult. You hear the words: that you are creating your reality moment by moment; that all you need to do is to love yourself, and none of you do, you know. That all you need do is live in joy, and you say 'well, I'm doing my best'. We understand, it is very difficult. And when indeed you do manifest diseasement of the body you fall into a black hole and judge the judgment which has created the diseasement in the first place. You have the idea of the perfect body, and you have the idea that if you are in perfect alignment, your body will be healthy. So, when you are not in perfect and vibrant health you judge yourselves for being so stupid, so un-aligned, so un-enlightened to create this diseasement. You say 'now everybody will know that I'm not an enlightened being'. We know - we hear you. You broadcast it loud and clear. So do you understand, all of you, how you do it to yourselves? It is very tricky, eh? And always you are worried that somebody will judge you as harshly as you judge yourself.

Q: So if it happens and one is living as joyful as one can, but it happens, one just acknowledges it and keeps going?

P: Indeed, but dear one, also: of course, you acknowledge the fear and you acknowledge the emotional pain which has created the dis-easement. It is to take responsibility indeed that you have created it. It is to align the judgment you have about yourself, about the diseasement, hm? It is to embrace. It is to know that all of it is Divine expression. All, ALL of it are opportunities for you to come into balance. But you create the change by the non-doing, you create the change by the embracement. You may take your dis-easement again as if it were the child you are. Take it into you arms, caress it and rock it and sing it a lullaby. Not to force it away. Not to try to suppress it out of existence, but to love it into the light.

Q: (M) Greetings, P'taah. I would like to ask you if you could say something about timing. Lately my own mind has been turning to timing; about when I should do something, or when other people should do something.

P'taah: Hm, indeed. Timing, hm? The time is right when it makes your heart sing. That is all. And as for everybody else's timing, well indeed, beloved, it is called their business. As you go in your day to day life, do not worry about the timing, beloved, because there is a perfect clock within your heart. As for 'doing it', forget it. When you 'do' what makes the heart sing it is called allowance of all possibilities to occur, and it is indeed in the knowing that it will all manifest itself for the greatest benefit of who you are. Now, you know what is the greatest impediment? *Doubt.* We have said to you before, *you* are the ones who create your reality. You put forth the thought into the universe and the whole universe rearranges itself according to what you believe about your universe and about yourself. When I say forth to you that you are powerful manifestors, that you may create anything you desire, then you say 'God/Goddess of my being I desire whatever it is, and I know it IS', and then you say 'Gee, I hope that works'. *(The audience is greatly amused.)* Well, you just screwed that one up. Have you noticed, dear ones, that when you casually put forth a thought, a desire of the heart - something quite trivial - and you just let it go and you do not think of it again and then a little time later, in your physical reality it occurs. You say to yourself: 'Hm, I was only thinking about that some months ago, desiring it, and here it is. How extraordinary'. You forget to say: 'How wonderful am I. I have created this.' You see beloved ones, that is how you create your reality in all the years of your life thus far, and every other lifetime you have had, and you do not even understand that you are doing it. So when you are doubting how powerful you are, look at your life. When you are looking at all the things you say are very 'shitty', you may also understand how you have done it. Because you see, dear ones, the universe does not judge, whether or not it is something that you love or something that you fear, as you put forth the thought according to your beliefs, so you manifest it. So it is entirely up to you where you focus your energy. You see there is no good or bad in the universe. There simply IS. It is you who judge if it be very good fun, or very 'shitty'.

Q: (F) P'taah, sometimes our minds wander, and we have thoughts that we do not really want to manifest. I am just not quite clear what the difference is: If I want something to happen, but I am not trying too hard, and then I have an unguarded thought that I definitely do not want to happen, why would one manifest and not the other or does it all manifest?

P'taah: It is called emotion. It is also called fear. It is the intensity of feeling.

Q: There is a difference between hanging on to something and wanting it to happen and having a free emotion that lets it happen.

P'taah: How extraordinary.

Q: I am not quite clear about that. I know I said it, but I am not quite clear.

P'taah: Well, there is for instance 'need'. Hm?

Q: So if something is really like: 'I need it', it is not just a useless thought, is that the difference?

P'taah: Have you ever noticed that when you think you are in dire straits and you need something by Friday, it does not occur? And you say 'If I am so powerful, how come it did not happen on Friday?'

Q: Yes.

P'taah: That is called need. Gods and Goddesses do not need, they only desire and create through that desire, not through the need, because you see, dear one, need is that face of the coin which is called desperation, called fear, very well clothed.

Q: Sometimes one gets so far that one cannot even be bothered to get in desperation any more. Sometimes one is so tired of desperation, that one thinks: well if it happens it happens and if it doesn't it doesn't. Often those things manifest as you...

P'taah: But of course, once you let go and allow any possibility you will create what it is you desire from your heart.

Q: Often, when we have no investment in that any more?

P'taah: Exactly. And we have said before that one of the inhibitors of your manifestation in physical reality is that you close down your possibilities and probabilities in the expectation of how and when it may occur.

Q: That is expectation and also, when we are putting pressure on how the universe shall act. When we allow it to do it in its own way, then it manifests.

P'taah: Indeed. And you know, the greatest creative part of you is in the knowing of what is also for your greatest benefit.

Q: My total consciousness knows that? So I only have to let go and let my greater consciousness take over, which means I remove myself from masculine, mental energy.

P'taah: Taking off your head and putting it under your arm.

Q: So that is what we, men and women alike, have resistance to. We think we are being irresponsible when we stop thinking, is that it? The big thing that we're all stuck on?

P'taah: Why are you asking? You know perfectly well. (Laughter.) But indeed that is how it is. *(P'taah focuses his attention on a gentleman, who joined the audience after the break.)*

Well, entity, how are you?

Q: Excellent, thank you. Really great.

P'taah: So you are, indeed. In physical reality, if you like. You have wondrous questions for us this evening, dear one?

241

Q: Not really. So many questions are being answered by just listening.

P'taah: How extraordinary.

Q: (F) P'taah, what is your reality, when you are not here in this room?

P'taah: Which one?

Q: I don't know.

P'taah: Beloved, there are myriad realities. We do not confine ourselves to one, and soon you will not either.

Q: That is good to hear.

P'taah: We will have more of this very soon, beloved. Come and visit us again for the next instalment of amazing revelations. *(Laughter.)*

Q: Okay, thanks.

Q: (M) P'taah, I know we can get to allowance ourselves, but could you give us some prose or something, so we can remember you, so we can reach our allowance - so we may travel in the multiverses with you.

P'taah: Beloved, *(very gently)* you do. You see, dear one, it is merely to know that you may, and I promise you: you may. It is merely before the state of your sleeping that you remind yourself, that you ask forth from the God/Goddess of your being that you may have conscious memory. And you may call forth to us, dear one, and we will come for you.

Q: (F) P'taah, recently I have done a bit of creation, I think for the lesson of it, and got terribly stuck in the judgment and now I am afraid I will have to keep repeating the lesson.

P'taah: But indeed, you have to keep creating it until you get through the judgment.

Q: I am aware of it now, so can I not keep rolling with it to clear it away?

P'taah: But of course. Beloved, you are such a harsh judge of who you are and you have such judgment of the judgment. That is the curse of the New Age, eh? The phantom's revenge, the curse of the 'whoofties'. *(Laughter.)* Judging the judgment.

Q: I am aware of what I am doing and the wish to allow it, but it has not happened.

P'taah: Why do you think it has not?

Q: Because I am still in judgment?

P'taah: Hm, and beloved, it is all so serious. You know one of the ways you may align the judgment and the judgment of the judgment, is to see how totally ridiculous it is and have a jolly good laugh. In the laughter you may align it all.

Q: (M) Could you speak to us about how feelings create our reality?

P'taah: Indeed. What you are in truth is feeling and what is feeling? It is energy. It is a vibratory frequency. That is what you are, a vibratory frequency, and the frequency will change according to your feelings. As we have said before, very often you do not really know what it is you believe. It is so much part of that which is persona, the personality Self, that you do not even understand that this is what it is, that creates your reality. Now, most of humanity, in truth, are not so familiar with feeling. They are very familiar with resistance, they are very familiar with the terror to feel, because you see, beloved, it is the terror of annihilation. And so most of you would do anything to escape from feeling. When you allow feeling, however you may judge it to be, in fact [it is to know that it] is really neutral energy. When it is allowed, it moves freely through the body and creates joy, creates unity, creates non-separation. When you judge how it may be, you create - very often - resistance, which is pain. Pain is not feeling. We have spoken forth much of this and you may read [it], that it may explain fully to you how it be. *You see beloved, that which is transmutation is to allow feeling without judgment* - to allow that which is neutral energy, that [which] sits within the solar plexus, to travel forth to the heart, to travel forth to the crown to create ecstasy.

Beloved ones. *(To the host:)* Thank you dear one. It is sufficient unto the time. *(To the assistant technician:)* Thank you, beloved one. *(To the hostess:)* Our thanks, dear one - and how are your chooks[1]? *(Laughter.) (P'taah seems to be completely aware of the on-going*

---

[1] Chook, an Australian jargon word for chicken.

243

battle the hostess has with one of her chickens, which persistently scratches away the mulch of her fruit trees, and with the rooster, who attacks her at feeding time.)

Q: Well, they are just reflecting me, I suppose.

P'taah: Indeed.

Q: How do I tell the rooster that I am one with him, one life with him?

P'taah: Oh, dear one, you should probably give him what is called a belt over the ear. *(Great laughter and applause among the audience.)* Sometimes they are very stubborn, these reflections we create. Hm, it is called joke, beloved. *(Another lady picks up the subject:)*

Q: But P'taah, isn't there validity in when we feel fed up with something, that we can act how we feel in the moment? Isn't that what you have been telling us to do?

P'taah: But of course, dear one. It is not so serious, you know.

(Then, turning back to the hostess:) It is all right, dear one, you may rush out and deliver a box to the ear, eh? *(And to the audience:)* You know we only say it, because she does love that which is called chook and rooster who treat her so badly, eh? And all of you very often want to rush out and box the ears of somebody, or something, hm? And if you could truly sit and laugh about it, and know that the frustration, the crossness is all alright and as you can allow it to be alright it is transformed, and then you may allow the laughter to bubble forth from your belly and you create the change. And some of you take it so seriously. It is the desperate scrabble after enlightenment. It is, dear ones, the desperate desire to come home to who you really are. When you can allow that you are already home you will change the face of all that you know in this consciousness, in this reality. You will truly know that merely to have the desire, without expectation, without doubt, then, indeed, you are already home. I love you all.

Q: We love you too.

P'taah: I know. My desire is that you may all look at who you are in your mirror and see the jewel, and see the reflection of God in every facet of who you are. Dear ones, there is no separation. You simply ARE, and in your ISNESS, indeed I honour you. Good evening.

Chapter 14

FOURTEENTH TRANSMISSION
Date: 27th of November, 1991.

P'taah: Good evening, dear ones. You are all well come indeed. You know that this evening is called an ending and also a new beginning. For thus we have completed many chapters, which is really one chapter. And indeed, we are still remaining in your service, dear ones. So we will continue in this fashion. Now, we will change format a little because we would ask first that there be query to come forth at this time, and then we shall speak to you. Called closing chapter, indeed.

Q: (F) P'taah, you said that helping us is a two way thing. I mean, we get a real lot from you, and I was just wondering what you get back from us.

P'taah: Joy, beloved. Joy indeed, and there is also that which is called learning experience, that we may join together in this fashion and thusly I am able to see who you are and how it is with you. Hm? But always, always it is a joy to be with humanity. It is always joy to see an expansion of consciousness. To see that what is coming forth will indeed enlighten your planet.

Q: (M) I would like to ask about the angel of gold. Is it correct that gold is a step-down frequency of pure divine love into materiality? Should it, in fact be in the hands of the healers and not the barterers?

P'taah: Now, that which is gold, hm, much treasured, is indeed a reflection of the All that Is, it is also to be thought of as a symbol. Now dear one, you say should it be in the hands of healers and not of merchants, hm? Is healer greater than merchant, beloved? Better?

Q: Sometimes more life-restorative. I was particularly meaning that it seems to have an affinity, that life frequencies flow back to their natural harmony. Specifically in the presence of gold, that it may assist the ionisation of the atomic structure.

P'taah: Indeed. It thusly is a tool. But as we have said before, beloved, the only tool that you need, in truth, is your heart, hm? And as we have also said before, healing is effected by the person in need of the healing, hm? So it is not to get too carried away with it all. You see, beloved, whatever it is you are looking at, whether it be crystal or gold, or any other of the jewelled substances upon your planet, of the minerals, it is to know that it is only a reflection in that particular family *(minerals)*. There is also the reflection of your nature of flora and fauna. And indeed beloved, there is also the grand reflection in humanity; because as gold is the reflection of the All that Is, what is it that you think you are? Do you understand? So it is not to put one above the other, *it is in truth the greatest lesson of the humanity at this time to know that ALL, all things are reflections of All That Is; that there is nothing which is not imbued with the divine essence*

Q: (M) Good evening, P'taah. You said that the only thing that separates me from myself, my true SELF, is my belief. Please, tell me what is the best way to expand my beliefs, to get to know my true SELF.

P'taah: Indeed. Now, this is a very good question, because it is all very well to have an intellectual understanding. How do you in a practical sense come to this understanding of what in fact you do believe. Now, we have said before that for many of you, you do not really know what you believe, because those belief structures are so much part of the persona that you are no more conscious of them than you are of the hair on your head, or lack of. Now, when you are in a quandary, that is, when you are manifesting in a fashion that is not in harmony for you, it would be very practical for you to take pencil and paper. Write a question, for instance, about abundance, (that is always a very good one for humanity, eh? Love affairs and abundance); or let us take love affairs. Sometimes I think this is creating more anguish in humanity than lack of abundance. Now, on the top of your paper write the word LOVE or the word MONEY, and you may draw a line down the centre of your paper. On the left side of your paper write down all of the things that you judge to be negative about money or about love, relationships, and on the right side write down all of the things that you truly feel to be positive. You will be amazed. And you see dear ones, it is also [to know] that what you judge to be negative, very often, you know you 'should not [do]', hm? Especially all of you wondrous 'New Age' people. It is like your judgements. You know you 'should not' have them, so you judge your judgements. You know you 'should not' be thinking in a negative fashion, and so you make even grander attempts to cover up negative thoughts. To cast them out, to sit on them. Repression and suppression. So in this fashion you may truly come into an understanding of what you really do believe. Having come into the understanding that what is negative you very often did not understand that you really believed, you may then understand that it is no longer necessary. *Not* that it is bad. Not that it has to be 'gotten rid of', suppressed or repressed, but merely that it is no longer necessary; that these belief structures do not serve you any longer. Then it is to say 'but it is *alright*'. To embrace it into who you are. To say 'That is alright. That was the thought of yesterday, and now we may allow a greater understanding of how it [could] be'. And in this fashion you embrace the belief of yesterday and expand into the greater consciousness of today. So you may do this with any facet of your life which is not serving you well.

Q: P'taah, one more question. I would like to know more about love.

P'taah: So would everybody in the room. And indeed, so would everybody on your wondrous planet, beloved. *(Very softly:)* What is it that I can say to you about love? I can say to you, without it you are fighting for survival, and with it you become grand and creative masters, reflecting the All That Is to all about you. But you see, beloved, *it is not about loving anybody else, it is about loving you.* Because: unless you can live in love and honour of SELF, unless you truly understand that who you are is the Source expressing in third density, in every facet, in every shade, in every way, in every colour - and that all of it is alright - then how indeed may you express that love outside of who you are? And that which is love of child shown forth, is a mirror of how it may be of Self to SELF. Do you understand, beloved?

Q: (M) P'taah, this man's son (referring to the man who asked the previous question), if I may say, has a brain tumour. It has been very difficult for the members of the family to comprehend the reason for this. How it may be removed and what the purpose is for the boy to generate this as his soul choice. I would really like some understanding of this for this family, please.

P'taah: You know we have spoken forth before, and really expressly about this and about this family. There has also been another question about a child in extremes, and everybody [involved is] not in understanding of how it can be. And it is that people are saying 'how is it that a child in its innocence may be punished in this fashion', and that all who come in contact, especially the family, are to come into an understanding of grief. We are saying to you, beloved ones, always, always it is to remind yourselves that this life is not a 'one shot' affair. All of you who come together in a family situation have been together for many, many lifetimes, and you have chosen to re-create each lifetime together, each one playing a different role for the experience, so that each of you, in what may be called role reversal, may come into an understanding of that which has not been embraced in other lifetimes. So with each incarnation the consciousness is expanded further and further. Now, in this time, *in these years of your time, the lessons to be learned are coming thick and fast.* You are calling forth extremes to come into embracement, to come into transmutation of agony into ecstasy, that you may not only enlighten who you are, but also for the enlightenment of all of humanity and indeed, assisting the goddess, your Earth, in her own changes. Now, it is to know that with each incarnation you have already, at soul level, come together to decide what roles you will play in your family: who will be mother, who will be father, who will be sibling and what gender you will take. Then you outline the broad spectrum game plan. This is not to say that you do not have moment by moment choice, because you do have a choice in how you perceive each moment. Now, it may be so, that a child, at soul level, decides that their experience and the experience with the family will only be for short years. That is, that the soul desires only to experience a certain time which it has not experienced in this fashion before.

Now, you know beloved, we have spoken forth to you about transmutation. We have said, how - in the allowance - you may transform everything. We have spoken to you that pain and anguish are resistance to feeling. You may truly be in allowance, and know that the pain is eons and eons of pain. It is not merely the pain of this life, of this situation. You have created it all for the embracement, to understand what is the jewel within; to understand the jewel within Self and the reflection of that jewel within all of your family. It is called love and allowance. So the time that you are experiencing may be jewel upon jewel, creating a diadem to light up every incarnation that you have had with each of your family of this moment. So, beloved, be in great tenderness, have every moment as full as possible, that you are creating joy, moment by moment, without any thought of that which be future. It is to know indeed that love, beloved, is never ending. Ties of love in a family, that you have created in eons of time will go forth. When one is translated from this density to another, indeed, you may find that those ties are stronger than they are in physical consciousness and reality. Do you understand? Indeed.

Q: (M) P'taah, how much control or communication do we have with our soul. In one instance you say that we have complete choice, moment by moment...

P'taah: We are speaking at this moment of 'conscious' choice.

Q: In the other instance you say that the child may have decided for only a short stay with that family... there seems to be a conflict.

P'taah: It is truly not conflict. It is only that you are not understanding that there is no separation of soul and consciousness, except in the intellect. Now, we have said before beloved: take off the head and put it under the arm, because, truly, in this fashion it does get in the way. Now, you are not the only one, beloved, to be perplexed by this. Your own conscious reality is governed ninety percent of the time, we would say, by your intellect. Now we have said that your intellect is really there to serve you, but what has happened in these times, lifetime after lifetime, [is that] you have become the servant of your intellect, as you have become the servant of your ego. This is not to say that there is anything wrong with either intellect or ego, it is just that they have grown a little big for their boots. When you can become more intuitive, more allowing of your day to day reality being guided by your heart, and by what ever it is that makes your heart sing, then you will find that you become more and more in tune with that which is called soul, Spirit. In this fashion you will be consciously knowing what it is that your soul desires for your greatest benefit. Then, it will be the heart's joy and delight - moment by moment - to use the intellect and the ego to fulfil the heart's desire. In this fashion there is no separation between consciousness and soul energy. Your soul, beloved ones, has great integrity, is the divine thread. Your soul *knows* that it is expression of Divinity. Your intellect has forgotten that it has anything to do with Divinity. Do you understand? You can read about it, beloved, eh?

Q: (M) Dear P'taah, coming back to the previous question, you mentioned that nothing is cast in stone, so I understand that the child or the person can change it [the illness].

P'taah: Indeed. There may be change. Now, do you recall, dear one, that we have said that in your desiring to manifest what ever it is that you want in your life, it is to put forth the thought, and we have spoken to you about limitation created by expectation, and the limitation created by belief structures. Now, there is also to know, beloved ones, that your soul knows far better than your mind what is going to be of the greatest benefit for you. So, it is to say 'this indeed is what I desire', but it is not to lust after the outcome. It is to know that as you desire it, whichever way it comes forth is alright. Do you understand? Now in the instance where there is a grown human, a beautiful man, who has come into the fullness of his physical body, and has created a diseasement, it is certainly that the dis-easement is caused by what is emotionally not embraced. There is within the physical body always a way that the dis-easement will manifest itself according to the area of non-embracement of emotion. Now, in desiring to free the body of the dis-easement you can go to the source [of the diseasement] and transmute the pain. At cellular level the body may heal itself. However, it is also to know that when the lesson is learned, at soul level you may desire to leave, because it is not necessary to be here any longer. Do you understand?

Q: I do. I also understand that if you learn this pearl of wisdom, the disease becomes superfluous.

P'taah: Indeed. It becomes ease-ment, not dis-easement.

Q: (F) I believe that we are the sum total of all our ancestors. Within our genealogy, or DNA, we have all the emotions, all the positives and negatives of all our ancestors. To me it seems that the only way we can be truly clear to make correct or aware decisions in our life is to clean away all this genealogy that lies within us.

P'taah: Beloved, understand you are all your ancestors. There is no difference, and it is all occurring, every incarnation you have ever had, is occurring at the same time. When you clean up this act, you clean up all.

Q: Well, what's a good way to clean it up?

P'taah: What is it you think that we have been speaking about all of these weeks, beloved?

Q: Yes I know, but is there some practical exercise, something specific, like putting a soul star above our heads and covering ourselves with light. Is there something specific?

P'taah: Instant enlightenment, beloved?

Q: Sort of. Quicker than this way, anyhow.

P'taah: Well you see, it can happen in the blink of an eye. All you need to do is to allow it.

Now, dear ones, you know in these weeks we have said forth many words. These words have all been spoken many, many times before. You heard it all and you have heard it lifetime after lifetime and you do not hear. You listen so desperately, and you are all wanting to get out there and do it, to become enlightened - to leave this planet, to know everything outside of your world. You are all so desperate to 'do.' We have said to you week by week: *there is nothing to do.* We have said to you all of the knowledge of the multiverses is within you. We would ask you: how many thousands of years have there been people telling you this and you have not heard? I have said to you every thing that you perceive outside of yourself is a reflection to you. That which you judge to be negative indeed is a reflection of who you are. That which you judge to be wondrously beautiful, beloved ones, is also a reflection of who you are. We have said to you that whatever it is in your life that you resist, that you judge, that you push away, that you repress or suppress, all of it will come back to you. Whatever it is you resist will persist and persist until you come into allowance, until you come into embracement, until you come into non-judgement. We have said to you that truly, every facet of who you are, every thought you have ever had in every lifetime, every action, every part of you, every cell of your body - is an expression of divinity. And truly we have said there is no judgement outside of you. There is no judgement in the universe, and who you are is God/ Goddess, All That Is, expressing in this third density of reality. You are multidimensional wondrous beings. You are all spiritual masters. You are all scrabbling around to be spiritual, but beloved ones, *that is who you are.* Spiritual, wondrous beings, expressing in this density of reality. You ask why I come? Beloved ones, you do not understand that it is joy to behold you; each one of you. You are all terrified that you will miss the boat, terrified that the transition will occur and you will be left behind. You all think you have to do something to be worthy of getting there, and you do not understand that, in truth, you already are there. Be in allowance of your limited perception. It is alright. And as you allow it to be alright, so the perception will become grander and grander. You are surrounded by your soul brothers of all dimensions of reality. Beings who love you, who love your planet. You think that individually you have no power to change yourselves, your planet, and your destiny as a race, a species. You forget, and think that the Earth herself is powerless, is victim, but you see, my dearest ones, that which is your Earth is a grand Goddess who is mighty and powerful. Each of you is like a star within the firmament, and each one of you light up the galaxies.

Beloved ones, you are on the greatest adventure. You have chosen a glorious path, and it is called *living.* It is called moment by moment to know, in truth, that you may create your reality and your future in every moment. Do not be chained to your past. Do not be chained to the past history of your species. It is for you to know that there is no limit. You are grander than you could ever imagine and everything on your planet is part of who you are. You are truly not separate from *any thing*, you are truly not separate from *any body* and there is truly no separation between you and the infinite expression of all things. In every moment you have a choice. In every moment you may choose *love* or you may choose *fear*. Beloved ones, it is not to judge the fear. It is a valid and divine expression of who you are. It is to embrace it into you. Allow it to be. Bring it into the light of your beingness, and each time you do, you illuminate the galaxies, you light up the multiverses, you change the cellular structure of your bodies and you ignite the crystal firmament of that which is called your brain.

We love you all so much. It is truly wonderment and joy to be with you in this fashion. It will not cease. If you desire it to be so, then, indeed, we will return for as long as you want. We are here to serve you, because we love you, and that which we desire for all of you, for every human, is merely that all come home to who you truly are. So be it.

Beautiful, indeed, you are. Well now, you are truly making wondrous colours in this room, indeed, right through the ceiling. Dear Ones - Good Evening.

ABOUT THE CHANNEL

JANI KING was born and educated in New Zealand. Her first contact with P'taah was near her forest home in 1947. The second conscious contact took place in 1961, when P'taah gave her information that seemed to make no sense to Jani until 1988, when contact was made with the channelled entity Saint Germaine. In the years from 1961 Jani worked in various countries as a singer, dancer, a wife, a sailor and traveller, in the broadcasting industry, in restaurants, *"in anything that was not boring"*. The only things at all unusual were fairly regular sightings of space craft *("which one didn't talk of in case they took you away in a straight jacket")*, and telepathic communications with whales and dolphins *("which one also didn't talk about for the above reason.")*. Jani arrived in Australia in 1980 and now lives in the tropical North of Queensland where, she says: *"..the reef and the rainforest keep a person sane, in spite of 'spooks' and space ships"*.

GOD I AM
From Tragic To Magic
By Peter O. Erbe

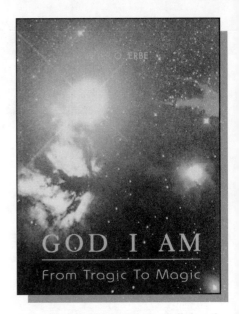

Every 25,000 years our Solar System completes one orbit around Alcione, the central sun of the Pleiades, a constellation at a distance of approximately 400 light years from our Sun. In 1961, science discovered a photon belt which encircles the Pleiades at a right angle to its orbital planes. Our Sun, and Earth with it, is entering this photon belt between now and the year 2011.

This photon belt is the cosmic 'trigger force' to shift Humanity from third level into fourth level density, from Separation into Oneness. Thus, the magnitude and beauty of this event Earth is preparing for defies any description. Earth and Humanity are aligning for its birth into Christ-consciousness - the union of Star Light with Matter - the marriage of Spirit with separated Selves.

As the night transforms into a new day, so is the Age of Darkness giving way to the Age of Light. It is the greatest event ever to grace the Earth and her children. Terms such as the New Age, Superconsciousness etc., are but different labels for one and the same occurrence.

It is the 'end-time' of the prophecies, for time as such shall cease to be. Ageing, ailments and sorrow shall be no more. To partake of this grandest of events, man must be aligned with its energy.

Humanity, as such, is governed by False Perception, the adherence to the frequency of Fear, the result of which is literally an upside-down perception of life. Only that which is aligned with Light can partake of Light, thus those not aligned with the cosmic current of

energy flow - the Divine Intent - shall sleep the long sleep.

It is the purpose of this material to develop the magnificent tool of True Perception, with which we align ourselves for the birth into the dawn of a new day in creation, the Age of Love.

As the chrysalis is the bridge between the caterpillar and the butterfly, so is True Perception the bridge between third level and fourth level density, between Separation and Oneness.

The universe with all its beings, in seen and unseen dimensions, joins with us in the greatest of all celebrations, the jubilance of rebirth into Light - the dance of the Gods - for where Earth, and we as her children go, is the fulfilment of the soul's ancient cry:

WE ARE COMING HOME

250 Pages - Available from your book store or write direct to:
IN AUSTRALIA: GEMCRAFT BOOKS, 291-293 Wattletree Rd., East Malvern, Vic. 3145
IN NEW ZEALAND: AQUARIAN BOOKS / LOTHIAN BOOKS
3/3 Marken Place, Glenfield, Auckland 10, Private Bag, Takapuna, N.Z.
IN U.S.A.: JEWELL MARKETING MARKETING SERVICES INC.
10400 Walrond Avenue, Kansas City, MO 64137 - Toll Free: 800 221 - 9183

TRIAD
PUBLISHERS PTY LTD